T0074022

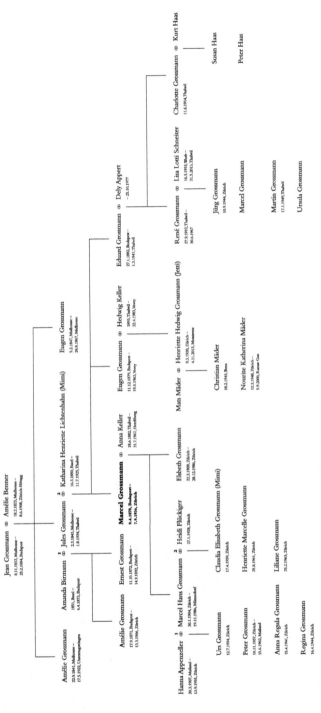

Graphics at the front of the book: Saskia Noll

Springer Biographies

More information about this series at http://www.springer.com/series/13617

Claudia Graf-Grossmann

Marcel Grossmann

For the Love of Mathematics

Claudia Graf-Grossmann
Grossmann Kommunikation
Schneisingen, Switzerland

Translated by William D. Brewer

The publishers and the author thank the following people and organizations for their support: Heidi Grossmann, Naturforschende Gesellschaft in Zürich NGZH and Fondazione Marco Besso, Rome

ISSN 2365-0613 ISSN 2365-0621 (electronic)
Springer Biographies
ISBN 978-3-319-90076-6 ISBN 978-3-319-90077-3 (eBook)
https://doi.org/10.1007/978-3-319-90077-3

Library of Congress Control Number: 2018940406

Printed on acid-free paper

This Springer imprint is published by the registered company Springer International Publishing AG part of Springer Nature
The registered company address is: Gewerbestrasse 11, 6330 Cham, Switzerland

For Christoph

Preface to the German Edition

A Bridge-Builder Between Physics and Mathematics

It gives me great pleasure to write a preface for this biography of Marcel Grossmann. The idea for this book arose during a visit of its author, Claudia Graf-Grossmann, and her sister to Pescara, the seat of ICRANet; and I am writing these lines shortly before the beginning of the fourteenth *Marcel Grossmann Meeting* which will take place in Rome during the 2015 International Year of Light, sponsored by UNESCO. This year also marks the 100th anniversary of Einstein's theory of General Relativity, as well as the Golden Jubilee of relativistic astrophysics.

What moved Abdus Salam and me to found the series of Marcel Grossmann Meetings in 1975? We were convinced that this European mathematician had built a bridge between mathematics and physics with far-reaching consequences, which right up to the present day can serve as a source of inspiration for modern scientists. In a book which is to appear shortly, *"Einstein, Fermi, Heisenberg and Relativistic Astrophysics: Personal Reflections"* [World Scientific, Singapore (in press, 2018)], I describe how the introduction of Special Relativity by Albert Einstein in 1905 placed our understanding of the fundamentals of physics on a completely new basis. His new approach, published under the title *"On the Electrodynamics of Moving Bodies"*, is based on a series of elementary observations, and on simple derivations using Euclidean geometry and linear partial differential equations which can be analytically integrated in a rather simple manner.

Here, the concept of space-time as a basic element for the formulation of all the laws of physics was established, and the brief article is still impressive today for its clarity, simplicity, and scientific relevance. And yet it remains difficult to understand Einstein's theory of Special Relativity, simply because it implies a change of paradigm.

When Einstein published the theory of General Relativity ten years later, it set entirely new standards: in an extremely subtle manner, it makes use of non-Euclidean Riemannian geometry, and its nonlinear Lagrangian description of the fundamental interactions leads to a set of nonlinear partial differential equations in four dimensions, for which, a hundred years later, still no general solution has been found. In particular, General Relativity leads to the revolutionary new fundamental idea that the gravitational interaction can be described geometrically by applying the equivalence principle. When Einstein was developing his theory, he could make use of the flat, four-dimensional non-Euclidean space-time which his former teacher in Zurich, Hermann Minkowski, had formulated for Special Relativity. Einstein transformed this approach into a curved, Riemannian geometric model of space-time, which behaves locally like Minkowski's space-time. He also made use of David Hilbert's understanding of the nonlinear Lagrangian theory and applied the most elegant instrument of higher mathematics and geometry of his time—absolute differential calculus—which had emerged from the Italian school of Luigi Bianchi, Gregorio Ricci-Curbastro and Tullio Levi-Civita.

The bridge between the physical insights of Albert Einstein and the theories of the Italian mathematicians was made possible through a close collaboration with his friend Marcel Grossmann. Like Special Relativity, General Relativity is difficult to understand, due to its enormous conceptual, mathematical and physical complexity. In addition, the latter offers few possibilities for concrete verification, in contrast to Special Relativity, whose direct connection to the Maxwell theory of electromagnetism makes it easy to test. It required an incomparable courage to formulate such a challenging theory, which was far removed from the predominant opinions on physics at the time and which pursued a revolutionary new approach to the geometric definition of physical interactions.

In Italy, Einstein's work had deep roots at the Institute for Mathematics in Rome, in particular thanks to the insights of Federigo Enriques, Guido Castelnuovo and, most especially, of Tullio Levi-Civita. Marcel Grossmann pointed out to Einstein the scientific thinking of Gregorio Ricci-Curbastro and of Levi-Civita concerning the absolute differential calculus. Einstein and Grossmann worked together in 1913 on a definition of the gravitational theory which would be based on this absolute differential calculus. Their joint publication (the "Outline" paper; cited in an appendix of this biography) includes two parts: the first, on the physical aspects of the theory, was written by Albert Einstein, and the second part, on its mathematical background, was written by Marcel Grossmann. Only recently have the various phases of the development of the Einstein–Grossmann equations come to light, when the unpublished notes of Michele Besso were submitted to a scientific historical analysis. These tedious and complex calculations aimed at explaining the anomalous precession of the perihelion of Mercury in terms of the new Einstein–Grossmann theory. The "Outline" published in 1913 did indeed not completely solve the problem, but the insights that it contained on the formulation of the equations and the classification of the astronomical data represented an important preparatory step for the later formulation and verification of the theory of General Relativity.

Before and after the publication of the theory, the contact between Einstein and Levi-Civita became increasingly close. Their cooperation led to mutual respect and to a friendship between the two men. I recall that Helen Dukas, Einstein's secretary of many years, showed me in Princeton the wonderful sentence that Einstein had written to Levi-Civita: "*I admire the elegance of your method of computation; it must be nice to ride through these fields upon the horse of true mathematics while the like of us have to make our way laboriously on foot*". There can be no doubt that the mathematical know-how of Tullio Levi-Civita, which was pointed out to Einstein by Marcel Grossmann, as well as Einstein's close association over many years to Michele Besso, with whom he could discuss all of his ideas, were fundamental for Albert Einstein on his path to attaining his goal of formulating the field equations of General Relativity—indeed, the most significant synthesis between physics and mathematics in the history of *homo sapiens*.

The Marcel Grossmann Meetings take place every three years in different cities: in Trieste, 1975 and 1979; Shanghai, 1982; Rome, 1985; Perth, 1988; Kyoto, 1991; Stanford, 1994; Jerusalem, 1997; Rome, 2000; Rio de Janeiro, 2003; Berlin, 2006; Paris, 2009; Stockholm, 2012; and, as already mentioned, again in Rome in 2015 and 2018. On the occasion of each Marcel Grossmann Meeting, awards are presented to individual scientists and to an institution (see also www.icra.it/mg/awards). The prizewinners are presented with a sculpture by the Italian artist Attilio Pierelli, whose original is made of silver and symbolizes the orbit of a particle around a black hole, after Kerr. For the first time in 2015, in addition to the MG Meeting itself, there were satellite meetings in Asia and South America.

We are happy to present this biography of Marcel Grossmann at a special ceremony in the *Palazzo Besso* in Rome. For, although Marcel Grossmann's scientific contribution to Albert Einstein's work is well known, there has been up to now little detailed information available on the man and the mathematician. This biography will fill that gap.

Rome, Italy

Prof. Remo Ruffini
Director of the International Center
for Relativistic Astrophysics Network (ICRANet);
Università di Roma 'La Sapienza'

Acknowledgements

Whoever delves as an author and a layperson into one of the most fascinating chapters of the history of science is skating on thin ice. Thanks to the support of the physicist and historian of science Prof. Tilman Sauer and of the mathematician Prof. (ret.) Dr. Urs Stammbach, I have avoided many perilous traps along the way. I accept the full responsibility for remaining notable *faux pas*! I wish to thank these two gentlemen heartily for their professional knowledge and their talent as detectives, for their attention to detail and for their patience. Tilman Sauer made incisive contributions towards the success of this biography and, with his scientific historical epilogue, he has lent it an essential gravity. Professor Dr. Remo J. Ruffini, theoretical physicist, founder and Director of the International Center for Relativistic Astrophysics, not only wrote the Preface to this book, but he has also made the name and the spirit of Marcel Grossmann known through the eponymous International Meetings, which he also founded.

Professor (ret.) Dr. József Illy went on a search for clues in the school archives of Budapest for this book. There, also, the art historian Edina Deme permitted me to obtain a fascinating insight into that city on the Danube around the turn of the last century. Barbara Wolff, from the Albert Einstein Archives at the Hebrew University of Jerusalem, was kind enough to critically read the manuscript, and she approved the publication of the 'Outline' of the theory of relativity in facsimile. Shaquona Crews of Princeton University Press gave permission to reproduce quotations, letters and photos of Albert Einstein. Daniel Schmutz of the *Bernischen Historischen Museum*, and the team of the ETH Library in Zurich, also gave me energetic support.

Furthermore, I wish to thank Dr. Hans Berger, the archivist of the Society of the *Constaffel*, Prof. (ret.) Dr. Martin Schwyzer, president of the *Naturforschenden Gesellschaft* in Zurich, and Dr. Hans Georg Schulthess for their valuable assistance. The *Confiserie Sprüngli* in Zurich researched their recipe archives, the *"Neue Zürcher Zeitung"* made possible many insights into the period described in the book, thanks to their digital archives, and the *Neue Helvetische Gesellschaft-Treffpunkt Schweiz* was so kind as to give me their book *'Kritischer Patriotismus'* by Catherine Guanzini and Peter Wegelin. The Municipal Archives of the cities of Basel, Schaffhausen and Zurich provided valuable background information. Anne Rüffer, the publisher at the *Römerhof Verlag*, as well as her associates Sandra Iseli, Saskia Noll und Selina Stuber, all helped the life of my grandfather achieve the form of an impressive book with their enthusiasm and professionality. I'm honoured that the English biography is being published by the prestigious Springer International Publishing AG and thank Dr. Angela Lahee for her commitment. Above all, I would like to thank Prof. William D. Brewer for his elegant and accurate translation.

Last but not least, I thank my family in England, Spain, Israel and Switzerland, who searched in old boxes and photo albums, gathered many interesting facts from the family history, and helped me by reading through the text. In particular, I am very grateful to my mother, Heidi Grossmann, for her notes on the illness of my grandfather and for her transcription of the diaries of my father, as well as for the travel notes on the walking tour of the two brothers Eugen and Marcel Grossmann. And I thank my husband, Christoph, for his loyal and tireless support from the beginning to the end of the two-year process of research and writing.

Contents

Prologue

In Zurich, a long, hot summer is coming to an end. Since June, the city on the Limmat has been suffering under the unusually high temperatures. The banks of Lake Zurich are besieged by young men with sun-tanned torsos, by girls in fashionable shorts, who lounge languorously on their beach towels and occasionally cool off in the waters of the lake. For older people and for people in poor health, in contrast, the weather has been a torment.

In the high-ceilinged rooms of the Paracelsus Clinic at *Seefeldquai* No. 49, near the Riesbach Harbour, the mercury has climbed to over 20 °C. even in the early morning, and the nuns have all they can do to cool hot foreheads and change bedsheets. They march tirelessly up and down the stairs in the romantic, impractical Italian-style building; their freshly starched habits soon look spotty and wrinkled, and the white bands of their caps are askew above the perspiration on their foreheads. But the women of the Menzinger Order are true to their reputation as hospital nurses: they radiate calmness and good humour and even manage to conjure up a smile now and then.

By 7 September 1936, the temperature has finally dropped to normal, and it is raining intermittently. In the evening, a light drizzle continues to fall onto the dried-up lawns and the prematurely coloured leaves of the trees in the garden. Does the patient, who is lying feverish in his bed and breathing heavily, hear the soft hissing of the raindrops and the rustling of the leaves? Or are the sounds of long-past times going through his head? The laughter of children, the scrunch of a sled gliding through a winter landscape, the jingle of the bells on the harness of a snorting pony?

Marcel Grossman has long since been living in the world of his memories and his thoughts; he can no longer stand up or speak clearly and is completely dependent on the help of others. Anna, his wife, understands him even without words. She sits beside him for long hours, sometimes working on her sewing; often, she falls asleep in her chair. The past months and years have taken their toll, and the continuing worries over the health of her husband, with the bedsores of the patient who is confined to his bed, have worn her down. Anna's arm causes her pain almost all of the time. But at least she can breathe freely now; others are taking care of Marcel and trying to relieve his bodily afflictions. At regular intervals, she turns over the almost motionless, heavy man. His bedding is changed; he is powdered and rubbed with salve. They give him sips of tea and try to lower his fever with cool wrappings. That is about all they can do to relieve the advanced pneumonia which has befallen his body, weakened by multiple sclerosis. Alexander Fleming has indeed already recognized the antibacterial action of penicillin, but it will be years before the first antibiotics, with their beneficial effects, are available to patients.

The atmosphere of quiet care that dominates the room spreads a feeling of calmness and peace. Slowly, the breast of the patient rises and sinks; at times, he gives a rattling cough. His breathing is pressed, distressed, and comes at longer and longer intervals.

1

A Spirit of New Beginnings

"Marcel was born on the 9th of April, 1878, an endearing and sturdy child, who was a great joy to us [1]." That is how Marcel Grossmann's father later remembered the birth of his son. A great joy—that is something that the young merchant values highly, given that years of an emotional roller-coaster ride lie behind him.

And yet, everything had begun so hopefully. Jules Grossmann is one of the enterprising young tradesmen and merchants who are active in the aspiring Danube monarchy of Hungary. He was born on March 2nd, 1843 in Mulhouse [Alsace], as the son of a Swiss family living in France—and, according to French law at the time, he is a French citizen. His father Jean had also been born in Alsace, whilst *his* father, Jules' grandfather, had left his home community of Höngg, near Zurich, as a fourteen-year-old. There, the Grossmann family was first mentioned in the rent-roll of the Einsiedeln Cloister from the year 1331, as farmers with their own land. Probably for economic reasons, Jules' grandfather emigrated as a teenager to the small Republic of Mulhouse, which was undergoing rapid industrial development and had been closely allied with Switzerland for centuries, to begin a new life there. Jules' parents, Jean and Amélie, operate a commercial firm there. The inventory is variable, mainly barrels and kegs (Amélie's family were bucket and tub makers), but also wooden and basketry articles, sometimes toys—to the great pleasure of Jules and his siblings, Amélie and Eugen. The children grow up bilingual; their mother tongue, French, remains ever-present for them.

© Springer International Publishing AG, part of Springer Nature 2018
C. Graf-Grossmann, *Marcel Grossmann*, Springer Biographies,
https://doi.org/10.1007/978-3-319-90077-3_1

Illustrations Jean Grossmann and Amélie Grossmann-Benner (oil painting, about 1840); and an advertisement for *Japy Frères*

After completing school and an apprenticeship as a merchant in Mulhouse, Jules is allowed to go to work for his uncle Albert Millot in Zurich as a trainee clerk (*commis*, as young office workers were called at the time). Millot is married to Jules' Aunt Adèle, had moved with her in 1852 from Mulhouse to Zurich, and from modest beginnings, he has earned respect and esteem in the miller's trade. He owns a factory for mill construction, in particular for roller mills, and deals in supplies for the needs of millers. Cast-iron objects, for example the elegant garden furnishings which charmed the bourgeoisie in the mid-19th century, round out his range of products. The duties of the young clerk are quite varied and carry a certain responsibility: He serves as cashier, receives deliveries of goods, sees to their storage and prepares the transport company papers, he writes invoices and purchases small apparatus and accessories. His work in Zurich is effective, and he could certainly have remained at his uncle's firm for a longer time. But he is a bright young man, sometimes a bit obstreperous and independent, and he brashly comes to the conclusion that his uncle's personnel policies are not to his liking. His decision is quickly taken: He will become his own Lord and Master. In 1867, he begins his new life of self-employment by visiting the World's Fair in Paris. He applies himself to dealing in supplies for the miller's trade, an occupation which he had learned in depth at his Uncle Millot's. On the way back, he visits the representative of a manufacturer of millstones in the French town of La Ferté-Chouard, and thus lays the ground for his future selection of commercial products.

The first domicile of the young company is a four-room apartment in the *Gerechtigkeitsgasse* No. 4 in Zurich. Jules Grossmann works and lives here. He takes a small bedroom for his own use; the remaining rooms serve as office and as storage space. The young start-up entrepreneur is confident; the spirit of the age fires his imagination. In his memoirs, he notes: "*Like the airy glimmer of the evening sun, I saw the passing of the olden times, whose history was long, great and even glorious, but at the same time I saw the beaming morning sun of the new epoch: Free trade and an impulsively expanding industrialism.*"

This pioneering mood was also predominant in Hungary which, following the revolution against the Hapsburgs in 1848/49, the Austrian defeat at Königgräz in 1866, and the Austro-Hungarian Compromise of 1867, had eked out a clear measure of independence from Austria. The administration of the country, which previously had been centrally located in Vienna, had been withdrawn. For merchants, this country is a veritable Eldorado, and in Switzerland the word is out that a promising new market is opening up in Hungary. That agricultural land was particularly interesting as a market for millers' equipment, and Jules had become aware of the multi-ethnic country whilst working for his Uncle Albert Millot. Now that he is independent, he travels several times to Hungary and establishes his first contacts to the mill owners there. Many of the Hungarian mills have Swiss directors or even owners; the managers, chief millers, chief machinists and office employees are also often Swiss.

The young man is impressed by the grandeur of *Pest-Buda*, as it was still called in those days, by the "*enormous, almost oriental hubbub*" in the splendid old State Railway station, and the "*melée of baggage carriers, omnibuses and other vehicles*". The picturesque location and the beauty of the city on the Danube with its castle, its majestic bridges, the wooded mountains in the West and its liveliness, "*not comparable to any Western European city*", capture his interest. As a sober businessman, he also senses that a market is opening up here, in which merchants can be successful even without knowing the Hungarian language. Although Magyar had been made the official state language in 1825, German is still spoken by a majority all over the country. Especially Swiss, who don't carry the 'stable aroma' of the Austrians, are welcome here. In 1870, according to the statistics of the Swiss consulate, 30 Swiss are working in the foodstuffs industry, of those 20 in the miller's trade. Of these, 12 are between 21 and 35 years young [2]. Between 1871 and 1891, the number of Swiss in Hungary will double from 516 to 1032, making them, after the Germans and the Italians, the third largest foreign group there [3].

Grossmann makes a great effort, contacting first the large and then the smaller mills. His success is indeed initially "*equal to zero*", but the young merchant perseveres and finally wins his first customers: He receives an unusually large order for millers' supplies, mainly silk gauze and millstones. The silk gauze is used for sieving the ground flour; gauze of varying mesh size is stretched onto wooden frames for that purpose. Stacked one above the other, they permit the separation of coarse meal, bran, fine flour and dust. The fragile gauze is cleaned regularly and has to be replaced at intervals. This is a line of business which promises repeated orders over a long period of time.

After Jules—aided by his sister Amélie—has initially operated his business from his apartment in the *Gerechtigkeitsgasse* in Zurich for a while, in 1869 he ventures the next great step and moves to *Pest-Buda*. The decision is not easy for him; he has become fond of Zurich (and of a certain young lady, whom he admires from afar when she attends the French church in the *Kirchgasse*). On the other hand, he is supported by his silk-gauze supplier, who wants to expand into the Hungarian market and promises orders. Wishing to be clear about his goals, Jules climbs up onto the 'home mountain' of Zurich, the *Üetliberg*, looks down on the city on the Limmat, considers the pros and cons—and comes to a clear decision, which sounds a bit precocious for such a young businessman, but is meant quite seriously: *les affaires avant tout* ["business before everything else"]. This motto will accompany him throughout his whole life. The die is cast; his move to the Danube is in preparation.

In 1869, Grossmann moves into his first business domicile in *Pest-Buda*, near the parish church on the banks of the Danube. His cousin Albert Hollender from Munich, with whom he has already undertaken a trip to Hungary and with whom he can work together quite well, likewise moves to the city, and will spend years at Jules' side working in the commercial firm. Grossmann later describes his impressions of Hungary at that time as follows: "*Up until 1867, Hungary was in fact a province of Austria, even though it was called the Kingdom of Hungary. Its territory included Slovakia in the North, Transylvania and Wallachia [Romania] in the East, in the South the Serbian Province and Slovenia, at its centre the large Magyar Province, and to the West, German-Hungarian regions. Great forests lie to the North, East and South. Enormous low-lying plains with the most fruitful lands, great deposits of coal and iron ore, rich vineyards and tremendous cattle-breeding areas. When travelling, one sees large herds of oxen, horned cattle, sheep and horses without end. The land is cut through by two great rivers, the Danube and the Tisza. The country's administration is extremely precarious. The Austrian administration has departed, and the Hungarian, often with an antiquated organisation, is not up to the task. [...] Industry is still spread very thinly. But the year 1867 brought a great opportunity for agriculture: There was a poor harvest in Western and Central Europe, whilst a record yield was obtained in Hungary. The sudden demand for cereals brought extraordinary profits for the country; steam mills, alcohol refineries, machine factories, banks and savings and loans sprang up so to speak out of nowhere.*"

The evolution of the population in the 19th century mirrors this stormy growth: The number of inhabitants of *Pest-Buda* increased from the end of the 18th century until the first years of the 20th from around 50,000 people to 900,000! [4].

In this flourishing land, young professionals and merchants feel quite at home. Jules Grossmann leads an active, gregarious life in *Pest-Buda*, working hard, but also spending happy hours within a circle of Swiss exiles of his own age. Engineers, architects, construction experts, merchants and machine technicians meet, and form a "*fresh, happy society*", which is soon organised into the "Swiss Support Association" (SUV). Grossmann, who owing to his birthplace in Mulhouse also has French citizenship, now renounces it in favour of Swiss citizenship, even though his heart will belong all his life to the *grande nation*.

The young Swiss have at their disposal a small, separate club room in the hotel "*Erzherzog Stefan*", where 'patriotic' newspapers are on offer and where they can play at games of cards. Grossmann is active for several years in the social section of the SUV, serves as cashier and makes many acquaintances. In 1870, at the largest steam mill of the time, the *Ersten Ofen-Pester-Dampfmühle*, which belongs to the successful Swiss industrialist *Haggenmacher* family, from Winterthur [5], he meets a "*pretty, slim young lady, whose aspect struck me quite strongly*". She is Amanda Birmann, the sister of the Swiss bookkeeper at the mill; Jules describes the meeting 50 years later as a *coup de foudre* [thunderclap]. What is this young woman, who grew up in Aarau in modest circumstances, doing in the city on the Danube? Did her parents send her to *Pest-Buda* in the hope that she would find a husband befitting her rank there? The idea seems not unlikely; there is a considerable excess of unmarried men among the Swiss exiles in Hungary. Whatever the goal of her trip may have been—Amanda Birmann is hard pressed, she is intimidated by the lively and noisy metropolis. She stays close to her brother, accompanies him to work and to events of the SUV. This is not hidden from Jules; he attends the social events more often than usual. During an excursion to the Johannesberg mountain (*Jánoshegy*), to the popular restaurant "*Zur schönen Schäferin*", he finally manages to strike up a conversation with her.

The young merchant is head over heels in love, but he at first hesitates—is he in a position to support a wife, a family? As he had already done in Zurich, Grossmann goes up onto a mountain in order to organise his thoughts: He climbs the many steps up to the terrace of the royal castle in *Ofen*, and once again, in view of the "*splendid city of Pest*", he considers what direction his path should take. In Zurich, his work gained the upper hand; this time it is *la voix du coeur* [the voice of the heart] that wins out. After his decision, he hesitates no more; evidently, neither do Amanda and her family, and within the same year, the wedding takes place. The bride is 19 years old, her groom eight years older. They must have made a handsome pair when they went strolling together in *Pest-Buda*, the dark Jules Grossmann with his dapper parted hair and his confident regard, giving friendly greetings to people on all sides; and his tall, slender, light-brown-haired wife with her alabaster complexion at his side.

These two people, with their such different dispositions, complemented each other well: Jules is confident, pragmatic and decisive, whilst Amanda is more introverted. She all the more enjoys furnishing in a homely manner the new dwelling of the young married couple, on the 2nd floor of a large apartment building in the Markos Street (*Markò utca*). She shows her good taste and aesthetic sense in the process.

In 1871, the first child arrives, Amélie. Father Jules is building up his firm, which has its headquarters just around the corner, on the ground floor at Markos St. and the Waitznerring (*Váci körút*, today called *Bajcsy-Zsilinszky út*). The location is quite advantageous for a commercial establishment: The elegant building, recently completed, is just 200 m from the Austrian State Railway station, on the broad boulevard which leads to the Basilica of St. Stephen. The mill quarter, somewhat further north on the banks of the Danube alongside the *Margareten* Island (*Margit Sziget*), is easily reachable, and deliveries can be quickly brought there.

Horse-drawn taxis pass by the house continually, and the first trolley car line is soon opened. Later, son Eugen describes the scene as follows: "*On this main street, there was a lively traffic of the horse-drawn trolley cars, some of which were double-deckers. Their horns could be heard all day long and late into the night. The many carriages and horse-drawn cabs which passed by in full gallop also caused a thunderous noise of hoof claps. And then sometimes the soldiers from the nearby barracks would march past, which was not always enjoyable for the olfactory sense, but, when the military band was playing, it was a treat to the ears. A highlight which we experienced several times was the passage of the Emperor Franz Josef in his yellow-black court carriage, arriving from the nearby West Station, saluting to either side to the tune of the enthusiastic* Eljen *("long live the emperor") called out by the crowd of spectators* [6]."

In the meantime, Jules Grossmann can also afford his own carriage, a modest one-horse buggy. He receives customers, intensifies his contacts tirelessly, and keeps his eyes open on the metropolis. Its aspect is changing rapidly: Whoever can afford it is building an elegant villa, a city residence, or an opulent company office. *Pest-Buda* is justifiably proud of its location on the Danube and wants to keep pace with the most beautiful cities of Europe. It is often called the "Paris of the East", and the developments on the Seine under Baron Haussmann are attentively followed and eagerly imitated. Buildings and palaces line the new generously-dimensioned and tree-bordered avenues and squares. Architects and contractors give their creativity free rein, planning and building in the neoclassical as well as an eclectic *Art Nouveau* style. This latter was especially favoured in Hungary; its people indulge in buildings from the French and Italian schools, and enrich those styles exuberantly with motives inspired by the Ottoman Empire.

Jules Grossmann notes with interest how luxuriously the facades are decorated, and remembers that during his apprenticeship at his uncle's, he had heard of a renowned French cast-iron foundry, *Val d'Osne*, near Paris. In June of 1871, he undertakes a trip to Paris so that he can form an opinion about the firm. He had postponed the trip several times due to the Franco-

Prussian war, with the siege of Paris and the uprising of the Paris Commune, but now, in early summer, a precarious calm had once again descended on the Seine. Bismarck has marched off following his victory, and the revolutionaries are in jail. Jules Grossmann experienced that at first hand: He visits the Palace of Versailles, where a few months before, the French-German peace treaty was signed, in which France ceded Alsace-Lorraine to the German Empire. In the *Orangerie*, Grossmann sees and hears the thousands of imprisoned *communards*, who are waiting there for their sentencing! These communards had incited a revolution against the conservative central government between March and May of 1871, and had unsuccessfully tried to force the adoption of an administration along socialistic lines.

It must have been painful for the Mulhouse native to stand at the place where the loss of his homeland had been signed and sealed. From a business point of view, the trip to Paris was however worth the effort: The *Val d'Osne* company as yet has no representative on the Danube, and Jules soon has his first successes in selling cast-iron parts for artistically decorated balconies, railings, garden fences and candelabras. Naturally, the owners of these splendid villas want to decorate their interiors in a corresponding manner, so the young merchant adds a product line from the Parisian firm of *Japy Frères* to his flourishing business; they are known for their high-quality iron goods, household articles and in particular for their splendid ornamental long-case and mantelpiece clocks.

Who in *Pest-Buda* can afford to commission the most famous architects and decorators for the design and construction of villas and company headquarters? Alongside Hungarians, Germans and Swiss, it is in particular the Jewish inhabitants of the city who contribute materially to the boom. On his arrival in the city, Jules Grossmann had been surprised at the number of representatives of the Jewish faith who are living there, primarily immigrants from Galicia. That region, which was divided up at the Yalta Conference in 1945 and now belongs to West Ukraine and to southern Poland, was formerly a part of the Austro-Hungarian empire, beginning in 1804. Large numbers of people moved out of the extremely poor agricultural regions and went to the cities in the West in search of work. And they found it there in quantity. In certain professions, the Jewish are predominant. Jules Grossmann attributes this to a *"remarkable lack of interest on the part of the Hungarian minor aristocracy"* in commerce and the liberal professions. The Jewish immigrants filled this gap, becoming successful lawyers and medical doctors, real estate agents, managers of health and vacation spas, detectives and police.

The Jewish community in *Pest-Buda* in Grossmann's time makes up nearly 25% of the city's population [7]. They are not only active and entrepreneurial, but also decidedly modern and open to new technologies, and often among the first to adopt them, as the telephone book from 1880 demonstrates. The economic growth which took place in those years had to be financed, or at least financially secured. This is frequently arranged by Jewish financiers. It is thus not out of place to say that the quantitative and qualitative period of development on the Danube could not have taken place without the Jewish community. This did not protect them from suffering the same fate as their fellow Jews in Germany beginning in the 1920's: Following the occupation of Hungary by the National Socialists in 1944, an *"uninhibited fury of deportations"* began, in the course of which within only a few weeks more than 400,000 Jews were sent off to Auschwitz [8].

In the 1870's, all of that still lay in the distant future; the mood in *Pest-Buda* was euphoric, and the trees seemed to be growing right up to the sky. Jules' and Amanda's family is also growing: In 1872, their second child, Ernest, is born. As was usual at the time, the Grossmanns hire a wet nurse to care for him, and she lives with them for a whole year. The baby thrives, but Amanda has been weakened by the second birth. And her psychological state has also not improved in the company of her sturdy husband; she remains timid and uncertain. When the head of the family undertakes an urgent business trip to Paris before Christmas of 1872, Amanda is frightfully depressed. By his own admission, the young father has lifelong regrets about this—in his own eyes unimportant—brief absence.

A year of extremes follows, and it plunges the young family into a deep crisis, which begins with the preparations for a great event: Vienna is making plans to host a World's Fair. It is the first time that such an event, which was organised twice each in London and in Paris beginning in 1851, is to take place in a German-speaking country. The Exposition opens with pomp and circumstance on May 1st, 1873. It is planned to be more grand than any previous event; the metropolis on the Danube is virtually turned inside out. The old fortifications are torn down and hauled away, and the *Ringstrasse* takes their place; the streets and the railway system are modernised, the Danube is channelled, and the Prater is beautified [9].

Jules Grossmann is kept busy simultaneously in the stands of his Parisian suppliers *Val d'Osne* and *Japy Frères* in the French section of the Exposition. The French want to prove to the world just how tasteful and artistic their products are. One imagines being in the cave of Ali Baba: Oversized canopy beds, period furniture, faience vases—and cast-iron ornaments, as far as the eye can see. Whole exotic villages are constructed; business is overflowing. Jules Grossmann takes care of customers and advises them, organises the sales operations of the stands, and arranges living quarters for company representatives in the overfilled city of Vienna, where 50,000 contributors and visitors to the Exposition are staying.

What a change in scene when he returns to Hungary: Amanda is sickly, nervous and short-tempered. An excursion to an afternoon concert at the *Redouten Saal*, which was intended to cheer her up, has to be interrupted because the young woman feels unwell. Her condition soon worsens; the doctors called to her bedside suspect typhoid fever. Her fever increases, her illness becomes more and more serious, and in April 1873, Amanda Grossmann-Birmann passes away at age 22. Her widower is, as he says, "*the most unhappy person on Earth*". Not only is he suddenly confronted with the necessity of taking care of the household and of two small children (rather unsuccessfully, as he himself admits), but also the loss of his first great love nearly bowls him over. His sister Amélie, who had already helped him during his time in Zurich, immediately travels to *Pest-Buda* to offer her support. She is indeed described as a splendid cook, but has essentially no experience with taking care of small children. Amélie Grossmann is spontaneous and decisive, but she has no special organisational talent. If this becomes obvious even to her brother, who knows little about matters of housekeeping, the chaos must have been hard to overlook. To add to the misery, a cholera epidemic breaks out in the city, owing to the poor sanitary conditions at the time and the rudimentary hygiene, without a sewage system. Jules travels back and forth between the Exposition in Vienna, with all its splendour and colourful business activity, and the silent and sad apartment at home in Hungary.

The joyless years pass slowly by. Jules Grossmann plunges into his work, which will be his support and refuge in the coming years. The children are taken care of by Amélie and a nurse, and slowly, a thin sheen of normality returns. But now the young father begins to feel unwell himself; he is overworked, often has headaches and is constantly tired. A period of rest and recuperation in Switzerland, in Brestenberg on the Hallwilersee in Aargau, restores him to health. There, the patients of the spa in the quaint castle under towering plane trees are not spared in their treatment: They are wrapped in steaming cloths and left to lie until perspiration flows from all their pores. Then they are doused with cold water, dried off and put under the rainwater shower. As part of the recuperation process, they have to drink lots of water and observe a strict diet. But evidently, many of them feel "*cheerful and joyous*" after this treatment, as claimed in an historical document [10].

Perhaps the voluntary and shared suffering often contributes here to the forming of tender relationships, as at all health spas. For Jules Grossmann, that is at first difficult, since he is still mourning his Amanda; but reason triumphs: He concludes that at 33, he is too young to be a permanent widower. He meets the 26-year old pastor's daughter Katharina Henriette Lichtenhahn from Basel, and in December 1876, at the Rhine Bend, they announce their engagement.

Henriette, called 'Mimi' at home, is descended from a family which traces its roots back to the 12th century, when they lived in and around Jena, in Thuringia, as Knights of Lichtenhayn [11]. Their descendants were mostly merchants; in 1524, the iron merchant Ludwig Liechtenhain became a citizen of Basel. Over the centuries, this family produced craftsmen, merchants, later also pastors, doctors, military men and artists. Henriette is not as beautiful as Amanda; she has a serious face and wears her hair parted and pulled back severely in a knot. But she has clear, shining eyes which look upon the world calmly and with confidence. Everyone who knew her describes her as a charming woman, warm-hearted, loving and blessed with humour. Jules and Henriette Grossman will spend nearly fifty years together. And what began as a practical marriage develops over time into a happy, harmonious and stable relationship; deep friendship and love unite these two persons, even though there is no more talk of a *coup de foudre*.

Following their wedding in 1877, Mimi moves to Budapest. Son Ernest recalls in his memoirs how their father prepares him, five years old at the time, and his six-year-old sister Amélie: "*It was evening, we were sitting on his knees and heard that he was planning a trip to Switzerland. In the way of children, we were quick with wishes, and were very surprised when we heard that he was going to bring us a new mother* [12]." Henriette takes on responsibility for the two little half-orphans quite naturally and with great energy, and rather soon there is once again a "*cosy, comfortable family life*". To be sure, she soon finds that the two children are not easy to deal with, as Ernest later self-critically admits. Their aunt had somewhat overdone the loving care and protection when she jumped in to help: Ernest and Amélie have never had contact to other children their own age and spend almost all their time in the apartment; they practically have to be dragged out to go on a walk. There is a confusion of languages in their little heads: They know some French, which they learned from a governess, speak Swiss German and know some Hungarian words. Here, the influence of their stepmother begins to have a beneficial effect. She has them take lessons in Hungarian from a private teacher, brings the two of them together with other children as often as possible, and makes sure that the two pale indoor children and bookworms get some fresh air. The hot, humid summers are spent by the Grossmanns, like other Swiss families, in a rented apartment in the country-side in *Leopoldifeld* (*Lipótmezö*), a ways outside of Budapest. That requires packing up all the furniture and household goods and shipping them off, which is gladly borne as the price for enjoying fresh country air during the summer months.

The year of Amanda's death, 1873, leaves lasting traces in economic and financial terms, also. Practically at the same time as the opening of the World's Fair, the financial centre Vienna was hit by a hefty economic and stock-market crash. It is thought to be due to a period of overheated eco-nomic development during the preceding years, which came to an abrupt end with the crash. Jules Grossmann describes the situation as follows: "*The effects spread out to Pest, to all of Austria-Hungary, to some extent also to other regions along the Danube and to Germany. The financial catastrophe paralysed credit lending and in particular the ongoing construction activities*". The con-struction boom in Budapest, as the city on the Danube has been called since 1873, when *Buda*, *Obuda*, and *Pest* were combined into a single metropol-itan area, has stagnated. Ornamental cast-iron products are no longer in demand, and the merchant is left holding his inventory. Grossmann liqui-dates his stocks, gets out of the cast-metal business, and concentrates on his original area, supplies for mills and agricultural operations. Then the tireless entrepreneur acquires a line of steam and locomobile threshing machines for his firm. In the meantime, he has several freight wagons and a stall of draft horses, which are quartered in the warehouse across from his apartment. He is happy to be able to concentrate on his work once more, whilst his new wife cares for the little family, and she is soon expecting a child. They find a larger apartment in the same building on the first floor at *Waitznerring* 76 (*Bajcsy-Zsilinszky út*). The house is still standing today, although the neigh-bourhood has changed drastically with the construction of the motorway and a number of new buildings around the station. In this house, on April 9th, 1878, under the sign of Aries, Marcel is born.

2

Childhood

Once again in the summer of 1878, the family leaves "*unbearably hot Pest*" and moves to the countryside. They spend their carefree summer days in Leopoldifeld, in the villa belonging to the director of the *Rochus* Municipal Hospital, Prof. Lajos Gebhardt. The villa is in a splendid location, and Henriette must have appreciated the fact that it was fully furnished. On the veranda, shaded by a ball acacia, little Marcel lies in his crib and is delighted by the motions of the leaves of the tree in the wind. At the end of the summer, they move back to the city, to their new apartment. It is handsome and practical, with four large high-ceilinged rooms. The children's room has a balcony at the corner, which is exciting for the small ones, as the traffic in this central location is "*enormous*". Here, the children of Jules' first marriage, Amélie and Ernest, receive instruction by a teacher from the German-Hungarian School on Déak Square, where later father Jules will also take lessons in the language of the land.

In the summer of 1879, Mimi visits her sister Ernestine Marti, taking little Marcel along. Her sister is the wife of a pastor in the Swiss town of Gurzelen, a village not far from Thun. In the neat chicken yard of the parsonage, the little boy crawls around chasing after the fowls, pulls himself up on the fence and undertakes his first awkward steps. The little chap makes it unmistakeably clear that he is no longer a baby. And indeed, Marcel does not for long remain the youngest of the family: Four months later, on the 11th of December, Eugen is born in Budapest during a period of "*extremely*

© Springer International Publishing AG, part of Springer Nature 2018
C. Graf-Grossmann, *Marcel Grossmann*, Springer Biographies,
https://doi.org/10.1007/978-3-319-90077-3_2

fierce cold"; he is the second child of Jules and Mimi Grossmann. The temperatures remain bitter cold, and the baby often has blue hands in spite of continuous heating of the apartment. But he thrives and squalls with all his might when he is hungry. This allows the nearly two-year-old Marcel to demonstrate his powers of observation: When their grandfather on their father's side is visiting from Mulhouse, and is standing in front of the nursery door, Marcel points to Eugen with his clumsy little hands and imitates him with a "*Queeh, queeh*".

Photos Henriette and Jules Grossmann (around 1880)

The icy temperatures finally give way to a mild spring. In the Grossmann family, sunshine also reigns. Mimi feels quite at home in Hungary and in the city, takes care of the four children and keeps up a circle of friends. The business develops splendidly; sales and profits increase from year to year, often by around 40%!

Photos (left): *Confiserie Gerbeaud*, Budapest. (right): Henriette, Marcel und Amélie Grossmann (1879)

To make more space for the growing inventory, Jules Grossmann rents a large plot of land across from his apartment, between the *Waitznerring* and the *Fabrikengasse* (*Györ utca*), where he builds showrooms and a warehouse. He hopes that his eldest son will later take over the business and tries to awaken his interest in a variety of ways. Ernest sits dutifully for several hours a day in the office copying addresses, and is bored with it. Father Jules is in the meantime industriously learning Hungarian. In 1881, he has made so much progress that he can read letters in Hungarian and understands their main points without difficulty. That is also necessary, since he is planning a sales campaign which would do justice to modern marketing methods: With diligence and persistence, he has acquired the land registries of all (!) of the districts in the 63 provinces of the big country, and is compiling a list of all the land owners of parcels of thirty *Joch* or more in each district (1 Hungarian *Joch* is approx. 1 acre). He will write to all of them and send a brochure which he has had printed in an edition of a million copies.

Jules Grossmann has acquired a solid reputation in Hungary as a dealer of agricultural machines. But it is becoming increasingly difficult to import the machines: a patriotic movement has arisen. The Hungarians want to show the neighbouring countries how strong and independent their nation is. The Danube monarchy is doing everything possible to transform itself from an importing country to a producing nation. The importation of products has been made more difficult by imposing customs fees and formalities, whilst companies that manufacture in Hungary are offered tax advantages. Grossmann decides to follow this trend and to produce his own machines in future. The thought is not particularly pleasant, since he is an experienced merchant, but not a manufacturer. But he has confidence in his abilities and believes that he can accomplish the transition to a manufacturing company. However, he can't manage it alone, since he has no experience in production. He considers various alternatives and asks his supplier of many years, the Rauschenbach Company in Schaffhausen, for a suggestion: How might a joint factory for threshing machines with hand and horse-powered operation be set up; which resources and capital would be required? Jules Grossmann has known Johannes Rauschenbach, the charismatic founder of the firm, since his early days at his uncle's company in Zurich; the Schaffhausen company is in the meantime one of the leading manufacturers of agricultural machinery in Switzerland, and Rauschenbach has an excellent reputation as an industrialist and politician. His company will later play an inglorious role in our story, but as yet, relations between the Rhine and the Danube are unclouded.

Photo Marcel, Ernest and Eugen Grossmann (from left to right; around 1882)

Whilst Jules Grossmann is spending his time thinking about expanding his company, an event is taking place in the House of Hapsburg which attracts an enormous amount of attention in Hungary as well as in Austria. The Austrian Crown Prince Rudolf, son of Emperor Franz Josef I and the Empress Elisabeth, the legendary 'Sissi', marries Crown Princess Stefanie of Belgium in 1881 in Vienna. Soon after the wedding, the couple pays a state visit to the Danube Monarchy. Around the Austrian Station, not far from the Grossmanns' apartment, one can see *"an unusually grandiose scene. The royal couple arrived with enormous pomp and circumstance; the aristocrats in their royal suite were rife with jewels and furs, all of them on horseback, and the Hungarians were in their picturesque livery, also on horseback. Marching bands, military formations and a festive crowd: an awesome picture"* recalls Jules Grossmann 50 years later. The Grossmanns, thanks to their balcony, have a box seat and look out over the sea of heads on the pavements. This colourful parade must have been among Marcel's first memories. The harmonious picture is however deceptive—the marriage of the next in line for the Austrian throne is unhappy, and Crown Prince Rudolf commits suicide, together with his lover, eight years later at the Mayerling Castle.

In August of that year, the tireless entrepreneur is in need of rest to recover from his strenuous business activities. He travels to Switzerland and undertakes a trek alone and on foot over the *Grimsel* and the *Oberalpstraße*. Afterwards, he takes the post carriage to Flims. In the hotel there, a communication from his supplier Rauschenbach reaches him, telling him that the project planning for the joint fabrication of threshing machines in Hungary is complete and ready to be looked over. Grossmann interrupts his vacation and travels without delay to Schaffhausen. There, he meets up with changed ownership, which however does not particularly disturb him. The founder of the firm, Johannes Rauschenbach, had died in March, but there is no reason to suspect that the company would be any less reliable following the death of its patron—Rauschenbach is a leading supplier of agricultural machines; what can go wrong?

There was no lack of warnings. Shortly after his return to Budapest, Jules is advised by an older acquaintance that *"young people seldom know when they should be cautious in expanding their business"*. This statement makes no particular impression on Grossmann, but it develops later that his friend was quite right. In November of 1881, the association contract is signed with the Rauschenbach company in Schaffhausen, and it will later prove to have devastating consequences for the family business in Budapest. The contract can be seen today in the Municipal Archives in Schaffhausen [13].

A month afterwards, Henriette and Jules Grossmann celebrate Christmas in the most cheerful of moods. The fearless Mimi decorates the tall

Christmas tree, standing on the top rung of the ladder, although she is eight months pregnant. On the 27th of January, 1882, son Eduard is born, the third child of Jules' second marriage.

Marcel Grossmann's childhood in Budapest is happy and untroubled. His mother Henriette takes energetic and warm-hearted care of her family, her employees, and their families. She helps found an agency which looks after young girls who are spending overseas exchange visits in Hungary. And she has a talent for handwork, and gives all those under her protection as well as countless godchildren gifts of handmade socks, jumpers and caps each year.

The Grossmanns' circle of friends keeps growing, although only a few Hungarian families belong to it. The differences in thinking and feeling between Hungarians and 'foreigners' are described as great. The Hungarian language, Magyar, is an obstacle to communication which not all the foreigners wish to overcome. It is much more comfortable to deal only with other German-speaking people. Mainly, these are Swiss, Germans from Saxony, or families from Bohemia and Romania who are German descendants and maintain contacts to each other. They meet up at the German school, the Lutheran church, or through business contacts. Jules and Mimi Grossmann are friends in particular with the Gaehler and Haggenmacher families, related by marriage. Grossmanns later rent an apartment from the German-Hungarian Schulek family. A certain Friedrich (or Fritz) Haller also belongs to their circle, a chief engineer for the Hungarian state railways, the socalled "k. and k. Austrian Railway Company" [k. and k. refers to *"kaiserlich und königlich"*, imperial and royal, since the Hapsburgs were Emperors of Austria and Kings of Hungary], a private enterprise financed with French capital. Haller is later to become the first director of the Swiss Office for Intellectual Property in Bern, and the friendship between him and Grossmann will even play an important role in the life of Albert Einstein. Another employee of the state railway is the Swiss mechanical engineer Fritz von Schulthess-Rechberg. He is somewhat younger than Jules and lives in Budapest with his growing family. This engineer, with a particular flair for mathematics—the story was told in his family that during his student days, he had *"known by heart"* 200 mathematical formulas [14]—greatly impresses the young Marcel, and may well have laid the cornerstone for his later love of mathematics.

In the circle of friends, and among the members of the Swiss Support Association, there is a happy and gregarious life, with many excursions and concerts for young and old. The Grossmann family continue to spend their summers in the country, in the villa rented from the Gebhardts in Leopoldifeld. Especially the suppers in the garden are unforgotten, when friends drive up onto the gravelled entryway with their teams, and the wife of the Swiss Consul Heinrich Haggenmacher chats amiably with Henriette at the tea table. The children enjoy these stays away from the city and have only pleasant memories of them later. Ernest describes the region as *"a charming valley, with extensive forests and small ranges of mountains"*. He discovers it in the company of his teacher, who has been hired at short notice for the summer to lure Ernest away from his books, and likewise lives at the villa. The two collect flowers, set up an herbarium, and Ernest discovers with surprise how interesting the natural sciences can be. Sometimes he also scares his younger brothers by telling them that wild animals and gypsies are running around in the woods surrounding Leopoldifeld.

In the evenings, the children run out to meet their father when he arrives home in his one-horse buggy from a day at his business. They are allowed to jump in and squeeze together happily in the buggy, or they climb up and sit next to the coachman on his high seat. Mucki, the tame draft horse, is the designated favourite of the younger set, and sometimes on Sunday mornings they try to ride him, mostly resulting in slipping off and landing on their seats in the dust.

The family employs not only a coachman, but also a cook, a maid, and a governess. Jules Grossmann has accomplished a degree of social advancement which would have been hard to imagine back in Switzerland. The salaries of the household employees are to be sure rather modest in Hungary in those times. But it is hard to overlook: Grossmanns are meanwhile numbered among the establishment in Budapest. This bourgeois, cultivated and comfortable atmosphere leaves its stamp on the childhood of the Grossmann siblings, and certainly explains why they all tend to lead a somewhat extravagant style of life as adults.

Father Jules places great value on an education appropriate to the family's standing, which also necessarily includes music: The older children receive piano and violin lessons, with only modest success. The mistress of the house, Henriette, who cooks well and with pleasure, learns a repertoire of Austro-Hungarian dishes with the aid of the cook, including spicy meat dishes, hearty pastries and delicate Viennese desserts. Eugen still recalls happily seventy years later the abundant Hungarian-Swiss cuisine of his mother, who puts meat on the table twice a day. Her desserts are occasionally inspired by the creations of her countryman, the Genevois Emile Gerbeaud. His elegant confectionery shop, with its marble counter tops and crystal chandeliers in the centre of Budapest, is still an Eldorado for gourmets today. One of his specialities, called *'Kuglerli'* after the confectioner Kugler, even found its way back to the Limmat, and is served at teatime there in the house of Marcel and Anna Grossmann.

3

Troubles

In 1883, operations are started up at the factory of the new firm 'Grossmann and Rauschenbach'. It is located on the outer *Waitznerstraße* (*Váci út*) near the Lehel Market. Rauschenbach delivers factory equipment and first examples of hand and horse-powered threshing machines, which are henceforth to be produced in Hungary. The foundry functions quite well from the outset, although the new building costs four times as much as originally projected. Rauschenbach provides a foundry master and a foreman, whilst Grossmann is responsible for setting up production of the remaining agricultural machines, which he had previously only sold.

Even though threshing machines were not such complicated constructions in those days, their production has its perils. One had to design a number of devices, construct them, test them and finally produce them in large numbers. As the newly-minted factory owner self-critically noted later on, this process is accompanied by *"enormous numbers of difficulties of technical, commercial and financial nature"*. In addition, the technology evolves rapidly, and the wishes of the customers turn from horse-powered machines to steam-driven machines. In the former, the driving force is produced by muscle power (mostly by horses walking in a circle), sometimes with water power, whilst the more modern technology of steam power opens up a realm of new possibilities.

© Springer International Publishing AG, part of Springer Nature 2018
C. Graf-Grossmann, *Marcel Grossmann*, Springer Biographies,
https://doi.org/10.1007/978-3-319-90077-3_3

Photo The Headquarters of the Grossmann and Rauschenbach Company, Budapest, were in this neighbourhood

The Hungarian mills and land owners rapidly turn to the new trend. Grossmann and Rauschenbach adapt to that and develop small steam engines from three horsepower upwards, but this new challenge places great stress on the firm. The machines delivered from Schaffhausen are not suitable for steam-powered drives and have to be rebuilt, causing a considerable rise in costs. Unfortunately, these unforeseen hurdles do not encourage the two business partners to work together more closely; instead, they put a strain on their relations. The firm is nominally led by Johannes Rauschenbach Jr., who however allows himself to be influenced by some of his managers and quickly loses patience. He accuses Jules Grossmann of having incorrectly estimated the market situation. Grossmann, for his part, accuses his partner in Schaffhausen of having miscalculated the costs and over-dimensioned the new building.

The years 1884 through 1889 are decidedly difficult for the enterprise. The initial capital is soon exhausted, but Rauschenbach is unwilling to provide any more funds. Grossmann, in contrast, puts in a lot of his own money, including loans from his relatives, and in particular tireless work for the company, in which he believes. In 1886, Johannes Rauschenbach makes the first suggestion to liquidate the firm in order to limit losses. Two years later, he repeats this demand, in a much more determined manner, and in 1889,

Rauschenbach comes to Budapest in person, intending to force his partner to give up. A glance at the balance sheet for 1889/90 shows that the production is making no profits, whilst financial write-offs are contributing to deep losses, and these are indeed blamed on Grossmann [15]. After a violent argument, the two business partners agree to set up a stock company which will guarantee Grossmann's continued work by contract over several years. Rauschenbach demands—and obtains—three-fifths of the stock issue. He has thus increased his share of the partnership from 51% in 1881 to 60%, an astounding manoeuvre for someone who claims not to believe in the company's future.

Rauschenbach makes good use of the fact that Grossmann has some difficulty in raising the money for his share of the stock. The business that he had built up over twenty years in Budapest is not enough to finance his stake; his private fortune and capital from his relatives have to be sacrificed as well. Jules Grossmann hopes in any case that now the foundation has been laid for a prosperous development of the firm. But the peace does not last long: Whilst the ink of the signatures on the new agreement has barely dried, Johannes Rauschenbach uses his comfortable stock majority to force his inconvenient partner out of the firm's management and to replace him with his own loyal vassals. Jules' oldest son Ernest remembers indignantly how his upright, somewhat naïve father is literally mobbed out of the company. The minutes of the General Meeting in 1890 tell the whole story [16]: Jules Grossmann, who after all was a co-founder of the company, is not mentioned at all. The management merely notes that "*one stockholder has decided to file suit against the validity of the resolutions of our general meeting of the past year and [...] this legal matter is at the stage of decision by a higher court*".

This development is very regrettable, as the new stock company is already showing a profit by the business year 1890/91, although the losses of the preceding years have not yet been made up. It is paradoxical that the two former partners, now that the firm has finally moved out of the loss zone, are spending their time, money and nerves in a bitter six-year-long legal dispute. Jules Grossmann shows himself to be extremely persistent, even stubborn. He won't give in, although his wife Henriette and many of his friends advise him to give up the battle and use his energy for other projects. But that would contradict his character. The merchant and manufacturer eventually wins his case (the corresponding documents are, significantly, not archived by Rauschenbach), but the final result is bitter: His business is gone, his private fortune has shrunk to a fraction of its original value owing to the loss of income and interest and to the legal costs, and his good reputation in Hungary is irreparably damaged. At least he manages to pay back the loans from his family without losses. He remains a stockholder in the firm and can sell the shares above par several years later. In 1929, the firm is producing and selling over 800 tractors and 1000 steam threshing machines annually. This is a late satisfaction to Jules, even though his psychological wounds and distress from the conflict will pain him for the rest of his life.

Photocopy of the Association Contract between Grossmann and Rauschenbach

Photos *Johannisgasse*, Budapest|Eugen Grossmann (1887)| and High School Building *Markosstraße*

A BUDAPESTI V. KERÜLETI

ÁLLAMI FŐREÁLISKOLÁNAK

TIZENNYOLCZADIK

ÉVI ÉRTESITŐJE.

AZ 188⁹/₉₀. TANÉV VÉGÉN

SZERKESZTÉ

HOFER KÁROLY,

KIRÁLYI IGAZGATÓ.

BUDAPEST, 1890.

LÉGRÁDY TESTVÉREK.

Photocopy of the title page of the yearbook of the High School (*Gymnasium*) *Markosstraße*, Budapest (1890)

-- 81 --

II. B) osztály.

A tanulók neve	Vallástan	Magyar nyelv	Német nyelv	Földrajz	Mennyiségtan	Természetrajz	Rajz. mértan	Szépirás	Tornázás	Magaviselet
1 Adler Gyula	1	2	1	1	1	1	1	2	2	1
Bassó Artur	3	3	2	3	3	3	3	3	3	1
Berger Hugó	3	3	3	3	3	3	3	3	3	2
Bloch Hugó	3	3	3	3	3	3	3	3	3	1
5 Bloch Frigyes	2	4	3	4	4	4	4	3	3	2
Bloch Mór	1	3	1	2	3	2	3	3	1	2
Brandl Artur	3	3	3	3	4	3	3	3	2	2
Brandl Vilmos	2	3	2	2	3	2	2	3	1	1
Braun Bódog	1	2	1	2	3	3	3	2	2	1
10 Csillag Ignácz	2	2	1	2	2	2	2	2	c. v. m.	1
Csillag Rezső	2	2	1	3	3	2	3	3	3	1
Dařilek Henrik	3	3	3	3	3	3	3	3	3	2
Fried Leó	1	2	1	2	1	1	2	2	3	1
Fried Náthán	1	2	2	3	3	3	3	3	3	2
15 Freund Bernát	3	3	2	3	2	3	2	3	2	2
Glaszner Ferencz	3	4	3	3	4	3	3	3	2	2
Goldstein Emil	3	3	2	3	3	2	3	3	2	2
Grolig Ernő	2	2	2	2	2	2	1	2	2	1
Groszmann Marczell	1	3	1	3	3	2	3	3 c. v.		2
20 Haasz Gyula	3	3	3	3	3	3	2	3	2	1
Hahn Ignácz	2	3	3	3	3	2	3	3	3	1
Hann Henrik	2	3	3	3	3	3	3	3	2	2
Heller Ödön	2	2	2	2	3	2	3	3	1	1
Holstein Samu	3	3	3	3	3	3	3	3	2	2
25 Jahn Oszkár	2	4	3	3	3	3	3	3	1	2
Janovits Lajos	3	3	2	3	4	2	3	3 c. v.		2
Karolusz Rikhárd	3	3	3	3	2	2	3	2	2	1
Kaufmann Jenő	2	2	3	2	2	2	2	2	2	1
Kerék Zoltán	3	3	4	3	4	3	3	3	2	3
30 Kirchlechner Emil	4	3	4	3	3	2	2	2	2	2
Klein Adolf	1	2	1	1	1	1	1	2	2	1
Kohn Henrik	1	2	2	2	2	2	2	2	2	1
Kohn Oszkár	1	2	2	1	2	2	2	2	1	1
Mangold Artur	3	2	2	3	3	2	3	2	2	1
35 Majdánszky István	2	3	1	2	2	2	3	2	1	1

6

Photocopy of the class list of the *Gymnasium Markosstraße* (extract)

Jules and Henriette Grossmann take pains to keep their business worries, and later their financial problems, away from their children. The young boys from the second marriage are by now going to school; Marcel begins elementary school in 1884 at the private institution of the German-Lutheran church, which had been established with the aid of his father. The class consists almost exclusively of German and Swiss children, who speak nearly no Hungarian. The teachers therefore give instruction at the beginning in High German, and only later do they begin to speak Magyar with the pupils.

Eugen, who goes to the same school as Marcel, has lively memories of the four teachers there: the friendly Mr. Luther, with his collection of toys with which he enlivens the first few anxious sessions for the new pupils; the somewhat more strict Mr. Alex; the kind-hearted, older Mr. Kurz; and finally the very strict Mr. Lux, who often uses his rattan stick for punishing those pupils who fail to meet his expectations, dashing it down on their backs. Marcel and Eugen learn quickly, demonstrate their rapid powers of comprehension and receive high marks.

In 1885, Ernest, now 13 years old, goes to live with his aunt and grandmother in Zurich. He has grown into a difficult young man, as he himself admits. His favourite word is '*but*'. This teenager has only a few friends and prefers to sit at home with his books; he is oversensitive and frequently rebellious. It is no wonder that his relations with his parents are often strained, although the younger boys greatly admire Ernest. Father Jules has again and again tried to interest Ernest in a career as a merchant in his own company, but all these efforts have failed miserably (just as they will do later with Marcel and Eugen). So continuing his education in Switzerland is a relief for both Ernest and for his parents, but a bitter loss for his younger brothers.

Two years later, the family travels to Switzerland; only their father remains in Hungary and makes use of the time to live out his motto, '*les affaires avant tout*'. In those times, that means in particular continuing the legal battle with Rauschenbach. Henriette is certainly not unhappy at the chance for a change of scene; her husband has become difficult, sitting silently at the family dinner table and gulping down his food without comment.

Although more and more dark clouds are gathering above the family business, they move again to a new apartment in Budapest. The old apartment on the *Waitznerstraße* is indeed centrally located, but the noise of the horse-drawn taxis and trolleys, with their clacking hooves and iron-rimmed wheels, has become too much for the Grossmanns; and in summer, the apartment is unbearably hot. In the cozy, tree-lined *Große Johannisgasse 7* (*Benczúr utca*), which runs parallel to the splendid avenue *Andrássy-út*, they find exactly what they want. Here, the Schulek family has a parcel of land with two buildings. One of them is an apartment house, containing a very roomy apartment with its own veranda and a view of the "*enormously big*" garden that lies between the houses on the *Große Johannisgasse* and the *Stadtwäldchenallee* (*Városligeti Fasor*). The lawns in the garden, cut through by walkways, cover nearly 2000 m², and the property is planted informally with hedges, berry bushes and trees; there is even a small exercise area for gymnastic practice. The owners, the Schulek family, live in a small wooden house with their numerous children. The offspring of both families are of about the same ages, and they romp together through the garden.

The nerve-wracking annual summer move to the freshness of the countryside in Leopoldifeld is now no longer necessary; the new apartment and the garden offer sufficiently cool hideaways even in the summer months. The boys and girls play croquet, build cabins in the woods and organise a customs border, where uniformed "customs officials" hold watch until late in the evening. Daily swims in the Danube make the summer heat more bearable. In the winter, the nearby ice field at the Zoological Garden is attractive for skating; sometimes, a military band even plays there.

Photos The garden gate at *Johannisgasse*, Budapest|The Matthias Church,
Budapest, reconstructed by Frigyes Schulek|In the garden at *Johannisgasse*,
Budapest, (seated): Fritz von Schulthess-Rechberg, Ernest, Jules and
Marcel Grossmann; (standing): [unknown], Eugen, Henriette, Amélie
and Eduard Grossmann (from left to right, around 1890)

Photocopy of the balance sheet of Grossmann and Rauschenbach (1885)

Bericht der Direction.

Geehrte Generalversammlung!

Wir begrüssen Sie bei unserer zweiten ordentlichen Generalversammlung und sind in der erfreulichen Lage, Ihnen von besseren Resultaten, als dies im letzten Jahre der Fall war, Mittheilung zu machen. Wir verdanken dieselben einestheils den diesjährigen im Allgemeinen gebesserten Absatz-Verhältnissen unserer Branche, anderentheils den vorgenommenen Reorganisationen in unserem Geschäfts-Betriebe.

Bei Anfertigung der Bilanz, welche wir die Ehre haben Ihnen heute vorzulegen, sind wir strenge nach den Vorschriften der Statuten vorgegangen.

Durch die nach Vorschrift des §. 38 der Statuten an unseren Baulichkeiten, Maschinen und Modellen vorgenommenen Abschreibungen, sowie durch die auf Grund eingehender Erhebungen gebotene Bewerthung unserer Aussenstände, erhöhte sich die mit fl. 79,501 vorgetragene Amortisation auf fl. 91,651 und die mit fl. 197,307·79 vorgetragenen Reserven auf fl. 207,339·07; endlich haben wir auch die restlichen Umschreibgebühren und Gründungskosten im Betrage von fl. 2627·20 bezahlt.

Der Geschäfts-Betrieb pro 1889/90 ergiebt hienach einen *Reingewinn von fl. 50,430·58,* welcher im Sinne des Gesetzes zur Verminderung des letztjährigen Verlustes von *fl. 251,062·67* zu verwenden ist, so dass ein Verlust-Vortrag von *fl. 200,632·09* auf neue Rechnung bleibt.

Wir bitten Sie nun die Bilanz zu prüfen und den Vortrag des obigen Verlust-Saldos auf neue Rechnung zum Beschlusse zu erheben und der Direction und dem Aufsichtsrathe das Absolutorium zu ertheilen.

Wir haben Sie ferner zu verständigen, dass ein Actionär gegen die Gültigkeit der Beschlüsse unserer vorjährigen Generalversammlung den Klageweg betreten hat und befindet sich diese Rechtssache im Stadium der obergerichtlichen Entscheidung.

Zu unserem aufrichtigem Bedauern hat sich das Mitglied unserer Direction Herr Heinrich Schmid-Schenk durch Domizilwechsel veranlasst gesehen seine Stelle zurückzulegen und sprechen wir ihm für seine Theilnahme an unseren Bemühungen unseren aufrichtigen Dank aus. Es wird Ihnen daher heute obliegen, einen Ersatz für denselben auf die Dauer eines Jahres zu wählen und falls Sie einverstanden sind die Anzahl der Directions-Mitglieder von 5 auf 6 zu erhöhen, noch ein weiteres Directions-Mitglied auf die Dauer eines Jahres zu wählen.

Photocopy of the Management Report of the General Meeting of the Grossmann and Rauschenbach company (1890)

Marcel, 10 years old, is undyingly enamoured of Helene von Schulthess-Rechberg, who is the same age as he. The two are inseparable. Eugen, for his part, has a crush on her younger sister Luise—a constellation which will repeat itself later on in the lives of the two brothers.

Frigyes Schulek, the owner of the property, is a drafting teacher at the Academy and an architect; he directed the reconstruction of the splendid neo-gothic Matthias Church in the historic district near the Castle. His greatest work is the monumental *Fischerbastei* [Fishers' Bastion], still one of the major tourist attractions in the city on the Danube. Schulek's wife is from Stuttgart; her maiden name was Rieke. She is not only lively, but can also be a veritable dragon. Eugen Grossmann recalls with shudders how she continually punished her children in a brutal fashion and occasionally even assaulted the neighbours' children. Evidently, their parents take no note of these little dramas in the shared garden. The Schuleks remain friends with the Grossmanns for a long time and later visit them in Switzerland.

An older governess lives with the Grossmanns, Ms. Patz, called "Patzi" by everyone and beloved of the children. She remains with the family for ten years, with interruptions. The sixty-year-old German-Bohemian knows many legends from her home country, and they make a great impression on the children. The endless forests with their fabled creatures, the White Lady who floats at night above the parapets and through the chambers of the castles, the devilish animal that lures travellers into the swamps—at bedtime, the three younger boys listen to Patzi's stories with a languorous creepiness. In contrast to Mrs. Schulek, she practices a rather lenient regime with the children. Her punishments consist of making the guilty party kneel or stand in the corner. Or, very sternly, she cancels dessert! Only in really serious cases do Henriette and Jules intervene. Mimi then forces herself to give a few soft blows to the torso, which is in the end rather amusing for everyone.

Many of the buildings from that period are still standing. Whilst the quiet, pleasant district was occupied mainly by middle-class families during the Grossmanns' time there, today the quarter is dominated by ambassadors' residences and representative villas. Between the *Benczúr utca* and the *Városligeti Fasor*, there are still some lavish gardens; the earliest plane trees are in the meantime 150 years old.

The winter of 1888/89 holds a special surprise for the three boys: Father Jules has acquired a pair of ponies and a carriage in the course of liquidating one of his agencies in Transylvania. Before he sells the team, he undertakes a wintry drive through the far-reaching landscape to Rákos with the boys. Marcel, Eugen and Ernst sit close to each other, warmly dressed and wrapped in blankets in the carriage which has been converted into a sleigh. The snow crunches under the hooves of the ponies, their breath rises up like the exhaust from a steam engine in the bone-chilling air, and the bells on the harness dance with the motion of the sleigh.

In the spring of 1889, Marcel finishes elementary school. He is now old enough for high school [*Gymnasium*], and his Hungarian is good enough for the State School. The *Dániel Berzseny Gymnasium* is located at Markos St. 18–20 (*Markó utca*), not far from their previous apartment. Here the pupils, although they are only 10 or 11 years old, are treated like college students. The teachers lecture for hours, without stopping to check a single time whether they are being understood. A major examination is given every few weeks, and Marcel does rather poorly. His marks improve only after his parents follow the example of most foreigners and engage a home teacher, who helps the children to master the material that they are learning in school. This advice stems from the oldest son, who visits his family at the end of April for the Easter holidays. The difficult teenager Ernest has developed into a thoughtful, empathetic young man, whose advice is gladly accepted by his parents. The younger boys are fascinated by their tall, slender brother with his unusually handsome, pale face, sitting nonchalantly at the garden table. The family travels together on a Danube steamer to *Visegrad*, the former Hungarian royal castle. During the excursion, Marcel begins to feel unwell. Back at home, his symptoms become more severe, and he now has a fever and abdominal cramps. Their family doctor, Dr. Samuel Weisz, diagnoses typhoid fever, which greatly upsets his parents. Especially in Jules Grossmann, this awakens worrisome recollections of his first wife's illness. But in the case of his son, the typhoid infection, often fatal in those times, proceeds quite moderately. Weisz notes approvingly on this occasion that Marcel has *"an excellent nervous disposition"*.

At Christmas, 1889, the mood is sadder than usual, since their father is depressed by his business worries. He indeed manages to maintain his family's standard of living, in spite of the considerable financial limitations which he is forced to put up with. But he knows best that they are living off their reserves, now that he has no income, but instead high costs for lawyers' fees, and those reserves are rapidly melting away. Added to that, he cannot leave the house during the holidays to pursue his beloved *affaires*. He sits around, broods over his files, and is often absent-minded and irritable during conversations with his family, as he himself admits. Naturally, this cannot be hidden from the children. They suffer from the pressure caused by the court case, which grinds on for six years and hangs over the family like a sword of Damocles. They often hardly dare to utter a word at the dinner table, for their father is inclined to angry outbursts that cause the whole house to tremble. A glimmer of light in the true sense of the word is the shared wish of the three boys to the Christ Child: A lantern. Despite the snowstorm, it is set up in the garden and solemnly inaugurated after the family has loaded the children onto a small sleigh belonging to the Schulek family (because of Eugen, who has injured his leg) and taken them to the Sunday School celebration and back.

In these difficult years, Henriette's character shows itself most clearly. Jules describes her as quiet, but always confident and cheerful. The pastor's daughter is a faithful Christian and has an influence on her husband, who in the years following Amanda's death has turned away from the church. With Mimi, he finds his way back to the faith, and it remains important to him for the rest of his life.

Henriette is the firm anchor of the Grossmann family. Not only do her weekly letters to Ernest prevent the shaky bridge to the rebellious oldest son from collapsing; she also always has time to listen to the children, mediates between them and her husband, awakens their curiosity and encourages their play. And she is unmatched in caring for the ill. In 1890, the son of their family doctor and friend Weisz becomes ill with whooping cough and is given into Mimi's care. She manages to care for the child all alone, without anyone in her family contracting the disease. How much this effort must have cost her in an apartment without electric light can only be guessed at by her husband: One morning, he counts over 30 matches which she had lit during the night in order to look after the sick child …

In early 1891, the whole family visits Ernest in Switzerland. The trip becomes an unplanned adventure. The difficulties begin on the first leg of the trip, to Vienna. A penetrating odour permeates their whole compartment on the train. All eyes are on Eduard, to whom sometimes "not very appetising things" still happen, in spite of his 9 years of age. But the mother's inspection of his trousers shows that Edi is innocent of causing the stink. The riddle is solved only when Mimi opens the well-garnished basket of food for the trip and takes out a roast chicken. Its flesh is crawling with maggots, and it is quickly jettisoned out the window!

During their short interval in Vienna, the family visits the park of the Emperor's palace. Edi, the youngest, complains of a nasty toothache, which seriously dampens their vacation mood. Finally, they are all sitting, exhausted, in the night express to Zurich. But there can be no question of a night's rest and a rapid train trip: A violent storm has caused interruptions in rail traffic, and the tracks are blocked in places by gravel slides. The train has to stop and wait whilst the tracks are cleared by torchlight. Parents and children have to change trains with their baggage for six people in the middle of the night, balancing on makeshift footbridges over the flooded section to the other train. In the end, the trip lasts 36 h, but Marcel and Eugen can hardly sit still out of pure curiosity and spend most of the time standing at the window of their compartment.

The family finally arrives in Zurich, very much behind schedule. They spend the first few days there with Aunt Amélie and Ernest, who acts as guide and shows his family the new Municipal Theatre (in modern times the Opera); he reports enthusiastically that he has managed to acquire a ticket to the première of "*Cavalleria Rusticana*" by Pietro Mascagni.

Grossmanns then travel on to Gurzelen, where they stay this time in the nearby guest house of a gentleman farmer. In the self-named "Grand Hotel Thalacker", there is much cheerful activity. The boys help on the farm, leading the cows to the drinking trough and learning how to use a scythe. And they develop enormous appetites for the wholesome Bernese cuisine. Years later, the thought of bacon with beans, the golden-brown *Rösti* [coarse-grated potatoes, roasted in a patty], and the Bernese *Strübli*, a kind of pastry made with pancake dough, will cause their mouths to water. Together with Henriette's Lichtenhahn relatives, who are lodging in the "*Haus Schlingenmoos*" belonging to the von Wattenwyl family, there are lively holiday goings-on. The "*Lichtenhahn humour celebrates orgies*" during parties in the woods: They put on plays that they have written themselves, organise sack races and eating competitions with sweet rolls.

The national holiday remains unforgotten: The occasion of the six-hundred-years commemoration of the Swiss Confederation, the 1st of August 1892, is celebrated by the whole nation for the first time. Alpine fires are lit in the high meadows, and a crackling bonfire is burning in the village, where Uncle (by marriage) Marti gives a speech. His devilish nephews eagerly await his getting stuck in the middle of his speech, but he fails to do them that "favour".

Father Jules encourages the older children to go on hikes and accompanies them on a climb to the Stockhorn. He is probably able to relax here for a few days and forget what is waiting for him on his return to Budapest: The legal battle with Rauschenbach is entering its next round.

The children, in contrast, are unrestrainedly eager to return to Budapest and see their friends there, the members of the "Banda". This is the name of the tribe that they have established, with a "constitution" written by Eugen, which they all follow with great seriousness. They even stage a "children's marriage": Eugen is the groom, the bride is Erika Schulek, the bridesmaid Margrit Gaehler, and Marcel serves as pastor. The "wedding" is followed by a grand ball. A high-class social event for the junior set!

Marcel spends most of his time during this period working on his home-built trolley. He and Eugen have set up an enormous network of tracks with many signals, different trolley lines and connections, using match boxes painted brown with numbers and signs as trolley cars. Marcel's whole room is filled with it, and in the summer months, it also operates in the garden. Following dinner, the trolley lines start up operation, with a scrupulously precise timetable.

Even more fascinating than the model railway, however, is the original; recently, the first electric trolleys have started their much-publicised operation in Budapest. The older boys are allowed to use it to go to school, which would take a half-hour by foot. In spite of the warnings of their parents, they can't be prevented from jumping off and on to "*the cars as they were hurtling along*", in Eugen's recollection. They are often chided in no uncertain terms by the conductor, which however does not at all dampen their untiring enthusiasm—quite the contrary. They pick up discarded tickets on the street and use these somewhat dirty but authentic passes for their own model trolley lines. Along with his model railway and his homework, Marcel busies himself during this period with a growing colony of rabbits in the garden.

The sons fortunately are not aware of it, but their carefree childhood is nearing its end. In the winter of 1891/92, it becomes apparent that they are having some difficulties in keeping up at their high school, in spite of supplementary lessons in the Hungarian language. Their older sister Amélie has finished Middle School and is now receiving instruction at home in French and English as well as in German literary history, swimming and dancing.

From Zurich, where brother Ernest is attending high school, a disturbing message arrives; a bulge in his aorta has been diagnosed. Is it due to the fact that he has grown too fast and already is more than 1.92 m tall? The doctors fear that it has led to a cardiac defect. Or is it due to a tumour in the lymphatic system pressing on the aorta? However the final diagnosis may turn out, the medical resources at the time are very limited; the doctors recommend rest and care. Ernest however doesn't take their advice very seriously; finally, after long additional years due to his moving and changing schools, graduation from high school is within his grasp. Although he has to prepare very seriously for the coming examinations, he visits his family in Budapest at Easter and seems to be happy and confident. Everyone hopes that after the final exams in the fall, he will have an opportunity to recover his health.

He also spends the summer vacation in Budapest, taking his younger brothers along to swimming lessons in the Danube, helping them with school problems, and he comes to the conclusion that Eugen—who has particular difficulties in learning Hungarian—is definitely qualified for high school level instruction. In contrast, he concludes that the home teacher is incompetent. Eugen thereupon transfers to the Calvinist Hungarian *Gymnasium, Lonyai-utezs*, where he immediately becomes and remains one of the best pupils. Marcel's marks also improve with the new home teacher.

The departure of the eldest son leaves everyone in a melancholy mood, he himself included, uncharacteristically. The train leaves the station and the waving family—but Ernest will never see them again. During the late summer, he writes his life story within a few weeks. The handwritten notes in his careful script suggest that he was a thoughtful and critical young person, who continues to read a great deal and devours the masterpieces of classical German literature. He is impressed by Lessing's *"Nathan der Weise"*, and by Dante's *"Inferno"* (there is also a Beatrice in Zurich, but he has not yet dared to speak to her). Ernest loves philosophy and is interested in theology, but he hates mathematics and physics, and is annoyed by narrow-minded teachers. The young man closes his brief, thoroughly critical reflections with an artistically decorated *finis*, and doesn't realise how close to the truth he has come with that word.

Jules Grossmann's recollections of the night of the 13th to the 14th of September, 1892, are still very precise fifty years later. He was extremely nervous and fearful, in a way that he has never before experienced and cannot explain; he can't sleep and is only able to get some rest in the early morning. In the later forenoon, a telegram arrives from his sister Amélie in Zurich, where the Gymnasiast is living: *"Ernest is very ill. Come immediately."* Whilst the father is preparing for the trip with the night train and is packing his suitcase, Marcel, Eugen and Eduard sit fearful and oppressed next to each other on the veranda. A short time later, the second telegraph messenger rings the bell, and soon after, the three brothers hear their mother and sister sobbing loudly. They run inside and decipher the telegram themselves: *"Ernest is dead. Heart failure."*

Ernest had taken the written final exams on September 14th, and the other members of his class turn up one after the other at the *"Häfelei"* at *Schoffelgasse* 11 in Zurich (today, it is the *"Alt Züri"* restaurant); it is the favourite hang-out of their high-school fraternity. They are all relieved and start making toasts. Ernest has just toasted his classmates and called out, *"Dir, Omar, meine Blume"* (probably a reference to Karl May's novel *"Erkämpftes Glück"* [*"Hard-won Happiness"*]), when he suffers a rupture of his aorta and collapses, to the horror of his comrades. The death of the young man just before his 20th birthday is another severe stroke of fate for the family.

The following winter is very cold and brings a great deal of snow to Hungary. Ernest's grandmother and his Aunt Amélie are visiting in Budapest, but even here, everything reminds them of his premature death. In February 1893, calamity also befalls their friends Fritz and Caroline von Schulthess-Rechberg: During an absence of the father of the family on a business trip, his wife becomes unexpectedly ill with pneumonia and dies [17]. For the widower and his seven children, Caroline's death is a catastrophe. Without hesitation, Henriette Grossmann takes the youngest child, five-month-old baby Henry, into her home.

All of the financial and private woes, the deaths and farewells in the family and in their circle of friends must have begun to depress even this tireless optimist; in addition, it is becoming more and more difficult to shield the children from these problems. The three younger boys are now 15, 14 and 11 years old—too young to completely understand the situation, but old enough to sense that their lives will soon change. Eugen later recalls that his *"sunny and joyful childhood would be followed by a period which cast shadows on my youth. [...] Already during the last four years of our stay in Budapest, the legal proceedings which were to rob my father of his fortune and his means of existence lay like a nightmare over us children. But even after our move to Switzerland, in Basel and later in Thalwil, financial troubles were continually visiting us and depressed our mood all the more since we had been accustomed earlier to a carefree life."*

In the end, Jules Grossmann obtains a certain degree of satisfaction in his dispute with Rauschenbach, and he sees part of his demands met. But his long-term business is lost and the greater part of his fortune used up. After the termination of this unfortunate chapter, he asks himself whether he should try to orient himself anew in Hungary, or return with his family to Switzerland. He plays with the thought of introducing the recently-available artificial fertilisers to Hungary. But due to his insufficient knowledge of chemistry and in consideration of the further instruction and education of his sons, he and his wife decide instead to leave the country. In October 1893, first all their belongings are loaded onto a removals wagon and then transported in a personally-sealed boxcar to Basel. When the family then boards the train in the West Station in Budapest, numerous friends and acquaintances come to bid them a hearty farewell. Former employees have come, some of whose children are godchildren of Henriette; nearly 50 people surround the travellers. Not to make the impression that they are financially ruined emigrés, the Grossmanns travel in a dignified style in a second-class compartment to Vienna ... and then continue on to Switzerland in the cheaper third class.

What must have been going through Jules Grossmann's head as the train set itself in motion, steaming and snorting? He is now 50 years old; he has spent half his life on the Danube. A quarter century in an aspiring nation which now stands immediately before a monumental event: In 1896, the millennium of the settlement of the land is celebrated with enormous effort, with national pride and pomp. This thousand-year celebration is in memory of the year 896, when Prince Árpád led the seven Magyar clans from which the Hungarians derive their origins into the Carpathian Basin. Everywhere, new buildings are being built, and older ones are renovated and beautified. The splendid Metro is constructed in Budapest, the second in Europe after London; the *Heldenplatz* ["Square of Heroes"] with the nearby fairytale castle provides a magnificent new sight, and along the wide axial streets, majestic buildings and palaces are erected. Budapest is one big construction site.

Jules Grossmann spent his happiest hours here, but he was also deeply despondent; he experienced unimagined economic success and yet met an ignominious end as a businessman. Does he think back to his anxious train trips between *Pest* and Vienna during the World's Fair, or does the image of his first love Amanda appear before his inner eye? Or are his thoughts, as always, directed firmly towards his next goal?

4

Beginning Again

Although Mimi and Jules Grossmann travel to Switzerland with mixed feelings, the family's departure is an exciting adventure for their sons. Marcel, Eugen and Eduard know their home country only from their sporadic visits during summer holidays and from stories told by their parents.

Upon arrival, they find Switzerland rather unsophisticated in comparison to Budapest. Even during previous visits, the boys had joked privately about the homely one-horse trolleys in Zurich; they are used to the fiery, rapidly trotting or even galloping two- or four-horse cars in Budapest. Here, they see no noble avenues, sometimes paved with industrial flooring so that the carriages raise less dust and make little noise. And they would search in vain for an electric trolley on the banks of the Limmat; the first one in Zurich will be put into service only the following year, in 1894.

But their initial impression is deceptive; progress has made its appearance in Switzerland, as well. The population has grown from 2,657,614 in 1870, within 20 years to 3,032,945 by 1890 [18], which represents an increase of 14%, after all. The country, poor in resources, is transforming itself from an agricultural to an industrial economy, with important impulses due to electrification and to steam power. It is the age of industrial pioneers such as Alfred Escher, Charles Brown and Walter Boveri, and of the bookbinder Hermann Greulich, who emigrated from Silesia and founded the Socialist Party in Switzerland in 1870. Many of the enterprising merchants, engineers and administrators are foreigners. Often, they have come to Switzerland to study, have taken root there, and now lay the groundwork for a political,

© Springer International Publishing AG, part of Springer Nature 2018
C. Graf-Grossmann, *Marcel Grossmann*, Springer Biographies,
https://doi.org/10.1007/978-3-319-90077-3_4

scientific and societal establishment which will leave its stamp on the following decade, including its social contrasts.

The railroad network, for a long time a stepchild in the young Confederation, is now being rapidly extended, and nowhere is this more noticeable than in Zurich. On the 29th of February, 1880, cannon shots from the height of the Polytechnic echo down onto the town, announcing the breakthrough of the Gotthard tunnel [19]; two years later, the tunnel is opened to traffic. And in terms of construction sites, Zurich need not shy away from a comparison with Budapest. Whole quarters of the city are being redesigned, and the narrow, closely-spaced Biedermeier houses along the medieval lanes are being demolished with no further ado to make way for broad streets with representative multi-storied buildings. The Spirit of Modernism is flowing along the Limmat also, even though understatement is more the style in the city of Zwingli than on the Danube.

A fresh wind is blowing not only through the cities; politically, innovations are following one another in rapid succession. After the complete revision of the Federal Constitution in 1874, the referendum is introduced, and this central pillar of direct democracy is complemented by popular initiatives in 1891. Consistent with that, education is being continually improved, so that the potential voters can properly exercise their new freedoms. Twenty years earlier, the National Council had approved the law founding a *"federal polytechnical college in connection with a school for higher studies of the exact, political and humanistic sciences"*—the cornerstone for the future ETH (*Eidgenössische Technische Hochschule*) [Swiss Federal Technical Institute] had been laid.

Alfred Escher, as a young high-level civil servant and member of the parliament of the Canton of Zurich, played a decisive role in establishing the Polytechnic Institute. After having recommended dispensing with federal institutions only a few years earlier, and instead making the University of Zurich *"still more federal"*, he has now changed his mind and thus paves the way for the new Institute [20].

But Jules Grossmann and his family are still far removed from thoughts of higher education. They are struggling with more immediate problems: Late autumn is never an ideal time to move to a new city, and that is no different in Basel in the year 1893. The family of six at first moves into a temporary apartment in the *Riehentorstraße* which was rented for them by Henriette's brother Hans Lichtenhahn. Christmas is around the corner, but the Grossmanns are not in a festive mood. Since their new home is only provisional, just the immediate necessities are unpacked from their suitcases.

Soon, again with the help of their brother-in-law, Jules and Mimi find a modest single-family row house with a small garden at *Florastraße* 21; they move there in early 1894 and transform it into a comfortable new domicile with their furniture from Budapest.

Although money is still in short supply, this new beginning stands under a lucky star. The widely branched family of the Lichtenhahn family takes the newcomers quickly under their wings and does everything to make their assimilation easier. Hans Lichtenhahn, who had already found the first apartment for them, proves to be a pillar of moral and practical support. He lives in the parsonage of *Klingental*, quite near the *Florastraße*. Here, on Sunday afternoons, the cousins of the Lichtenhahn and Grossmann families meet. Marcel and Eugen note with some regret that there is a clearcut excess of boys in the Lichtenhahn family. In Budapest, there were many girls in their circle of friends, and now that they have entered puberty, their interest in the opposite sex has intensified.

Photo *Florastraße*, Basel

For "*puritanical and financial reasons*", Jules and Mimi don't allow their sons to take dancing lessons or to meet girls of their own age at dances. However, they have nothing against school trips and excursions with relatives and friends, and the teenagers discover Basel and its surroundings, from

Baselbiet through the Markgräfler Land and on to Alsace. They make a close acquaintance with the fiery Markgräfler wine and become tipsy for the first time in their lives from it. Father Jules takes a trip with Marcel and Eugen to Mulhouse, shows them his native town and the house where he was born; and later, they travel to the garrison town of Belfort. They catch their first glimpse of French soldiers, hear with a pleasurable shiver the clarion trumpet tones of the military march *Sambre et Meuse*, and enjoy the French cuisine. Their father forgets his frugality for once and treats them to a delicious and enormous meal in the four-star hotel *Tonneau d'Or*. After five courses, they can't eat another bite, and finish off their festive lunch with coffee and cognac in a café.

Eugen later recalls the years he spent in Basel as the happiest of his life, and Marcel no doubt felt the same. These two, who had grown up in a large city, indeed continue to feel that Basel is somewhat small-scale and provincial, but they are enthusiastic about its diverse and substantial cultural offerings. They discover the museums, admire antique furniture, weapons, musical instruments and costumes in the Historical Museum in the *Barfüsserkirche*, go to concerts in the Basel Cathedral, and buy theatre tickets with their scanty allowances. Henriette Grossmann is more than happy to be back in Basel, even though she never complained openly of being homesick.

Starting off the three teenagers in school in their new home is unproblematic; which, considering the different school systems and languages, is anything but a matter of course. Marcel goes to the Higher Middle School [*Obere Realschule*] on the *Münsterplatz*, and Eugen to the Advanced High School, whilst Eduard, in order to improve his German, is put back into the top form of elementary school. The oldest son proves to be a model pupil, as is shown by his report card from April, 1894, after just five months at the new school [21]. He receives the best mark of 6 on a scale of 1 to 6 in all of his subjects except technical drawing, free-hand drawing, and gymnastics; this suggests that his previous instruction in Budapest was of a high quality. A similar result is seen in his School Leaving Certificate in the fall of 1896: Best marks for both diligence and achievement in German, French, history, geography, natural history, physics, chemistry, arithmetic and algebra, geometry, descriptive geometry, mechanics and technical drawing; "only" a 5 in English and free-hand drawing, and a 4 in gymnastics [22]. These results make him one of the two top graduates of his class. Intellectual work apparently continues to be more his strength than broad jumping and sporting exercises. With this brilliant certificate in his pocket, he departs the Higher Middle School. He is not yet aware of it, but he will later return here and himself stand before a class and give instruction; and, after another decade, he will carry on a fencing match with the school administrators. But for the present, he turns his back on Basel. The city on the Limmat, Zurich, is awaiting him, and there he begins a new and fascinating chapter of his life, as a student of mathematics.

Why mathematics, of all things? It is considered to be a cumbersome, particularly demanding discipline. Why and how does he overcome the wishes of his father, who had wanted an engineer in the family, and not a teacher, if his son is not at all willing to become a merchant? It would appear that Father Jules was simply not consulted; Marcel presented him with a *fait accompli*. Evidently his instruction in Budapest and Basel had awakened Marcel's interest in a science that is rich in traditions and nonetheless modern. Pythagoras, Euclid, Aristotle, René Descartes, Isaac Newton, Leonhard Euler: already in earliest Antiquity and still today, men like these continue to try to understand the Universe with the aid of mathematics. Marcel could have studied in Basel, but instead he decides to go to Zurich, to the Polytechnic Institute. No doubt its excellent reputation as a school for technologists attracts him; a part of him is actually an engineer after all.

Now, having been accepted as a student, he has to make some practical arrangements. Where will he live, should he commute by train from Basel, or search for housing in the neighbourhood of the Polytechnic? Just below the new College Quarter, which borders idyllically on the vineyards above the city, there are several private boarding houses, in which older, single ladies, mainly widows, rent out rooms for a modest price. One of them, Mrs. Hardmeyer, is an acquaintance of his mother's, and the beginning student takes up residence in one of her rooms at the start of the semester. He is no doubt a congenial boarder; he is modest and well brought-up. The young man has every interest in not making a poor impression, since his mother would be immediately informed of any unpleasant incidents.

Marcel Grossmann has developed from a bumbling youth into an exceedingly presentable young man. He wears snow-white shirts with high collars, which his mother lovingly washes, starches and irons on the weekends. His thick, wavy dark hair is combed back properly, and the somewhat narrow-set eyes above his prominent nose and sensual lips generally adopt a serious expression. This gravity however is deceptive; roguishness and a grating humour lie just below the surface, as his fellow students soon learn. From Monday to Friday, Marcel lives in Zurich, and on the weekends he travels to Basel, where his parents live in their row house with Amélie, Eugen and Edi; in contrast to Budapest, now lacking servants and their own carriage.

This change of social status cannot have been easy for Jules and Henriette Grossmann, although they do not complain. Mimi is to be sure very happy to be again residing in her home town and near to her family; how splendid to live almost in view of her brother's house! But it is undeniable that these two returnees, who had lived an upper-class life for many years, now have to turn over every penny twice. The former industrialist has indeed obtained a certain satisfaction in the Rauschenbach affair, but *"the cheating and the lies"* remain, not atoned. With what remains of the formerly proud fortune—at the height of his wealth in Budapest he was, by today's standards, a multi-millionaire—he now sets about building up a new business.

Whilst Jules is looking around in Basel and searching for new possibilities for a business venture, Henriette, Amélie and the younger sons are adjusting to life on the Rhine Bend. The daughter attends a Lyceum and plans to become a teacher. She is good at languages, but the natural sciences, and especially mathematics, cause difficulties for her, so that Marcel gives her tutoring on the weekends. In the end, Amélie succeeds in entering the seminar for teacher candidates. Now, however, it is her health which upsets her plans; she was always pale, like her mother Amanda and her brother Ernest, but now the doctors find that she is *"highly anemic"*. With considerable regrets, she leaves the seminar, and to console her and improve her health, her parents take her for a stay on the Atlantic coast, in Beuzeval (Houlgate), in Normandy.

Although Edi is still in school and plays 'cops and robbers' with children of the same age in the neighbourhood, the two older brothers both start to go their own ways. Marcel has come of age and enjoys his new-found freedoms to the hilt. Around 1896, he and Eugen take a strenuous and not unproblematic walking tour from Basel to Zermatt and Gurzelen, then back to Basel, which is described in detail in Eugen's diary. It reveals the two young men to be wide awake, interested in everything and with a lust for life—and sometimes also youthfully arrogant and prejudiced:

"The train that was to take us to Lucerne steamed in. The steps of the few travellers who were willing to depart at such an early hour echoed on the platform. The conductor closed the doors, a shrill whistle sounded, and the long row of cars started to move off. [...] The chuffing of the locomotive became more and more a wheezing, the more we climbed up into the higher regions of the Jura—suddenly, a whistle blast signalled our entry into the Hauenstein tunnel. I immediately made good use of the total darkness in our car to take a big gulp from our canteen, which was filled with wine." Thus fortified, the young hikers arrive at the *Vierwaldstättersee*. *"Lucerne has at present only a temporary*

railroad station. The new station is under construction, and with its dome-like roof, it makes more the impression of a theatre than of a station. [...] We turned our steps towards the harbour. Here, we met up with something that is inevitable on a trip in Switzerland, namely a great horde of Englishmen; from all sides their jargon rattled in our ears. Many nations were represented by those waiting for the steamer to arrive, but none of the others made such a stuffy impression as these sons of Albion." After this unflattering judgement, the brothers board the "City of Lucerne" and enjoy the voyage and the lovely weather.

In Flüelen, public transport comes to an end; now they begin hiking and are continually overtaken by hotel coaches which are bringing the passengers from the steamer comfortably to their hotels in Altdorf. Marcel and Eugen arrive in Amsteg and take a room in the "Pension Freihof", which *"for a price of Fr. 3 was thoroughly comfortable and pleasantly furnished. We put down our baggage and washed the dust off our faces. Then we sat in the pleasantly-situated garden, ordered a beer and took a leisurely rest from our efforts. The most splendid, bracing mountain air surrounded us, and from the river bed, a cooling spray rose up from the thunderously roaring Reuss, flowing over gravel and stones."* The next day, the two hike onwards to Andermatt; underway, they cook *"our soup with the help of Maggi's Soup Rolls"* and *"radically eat up"* the provisions that they had brought with them. Continuing along, they meet up with numerous Italians, who are underway on foot from Italy, with their belongings in red bundles and a loaf of bread, on the way to the German regions of Switzerland. The factories that are springing up all over there are attracting workers from the South.

Eugen and Marcel pass through the *Schöllenen* and the Reuss Gorge, then cross over the *Teufelsbrücke* [Devil's bridge]. Eugen is impressed by how *"the Reuss plunges down, foaming, through a narrow canyon; the whole quantity of water is a mass of foam which crashes onto the stones that form the river bed with a fearsome violence"*. They spend the night in Andermatt, but they sleep badly, because the military base there is overflowing with officers and men; in front of every door in the hotel, sabres, uniform coats, military helmets and bags are lying around. During the night, *"suddenly, trumpet notes sound out—they are blowing the tattoo, and several trumpeters are going from house to house through the whole town"*. Tired, the two brothers tackle the next strenuous stage of their trip, from Andermatt to Hospenthal, over the Furka Pass to Gletsch and on to Oberwald, on the following day. Their path leads sometimes through swampy regions and through mountainous areas *"with Alpine roses and Gentians, whole meadows were shimmering in red and blue"*. Shortly before Gletsch, they are caught for the first time in a pouring rain.

On arriving in Oberwald, they look "*as though we had come directly out of the Rhône*". The two young men take a room in the Hotel Furka and "*had to go* nolens volens *to bed, since we had not brought a change of clothes with us*". Whilst they are snug in bed, their clothes are drying in the kitchen. "*Now, the evening bell-ringing began, which was of a highly bizarre nature. At first, only a single bell was rung, by simply hammering on the bell with an iron tool of some sort; then, after a few minutes of silence, the* spectacle vulgo *started up again with another bell*". The next morning, their clothes are dry again, but one of the straps of their knapsack breaks; its leather had become brittle as a result of getting wet. This mishap did not exactly contribute to improving their mood. Nevertheless, Zermatt is calling, and so they press on through Niederwald and Fiesch. It begins to rain again, and they have to pass the time in a café. Since continuing on is out of the question, they explore the village after dinner during a pause in the rain, and observe with amazement the arrival of the Furka Post coach: "*The travellers get out of the coach whilst the horses are being changed. For this, no fewer than 14 nags were necessary; the main coach had a team of 5 horses, and each of the 3 trailers had another three.*" The stables of the two hotels in town are correspondingly spacious.

They set out in the morning on the next stage of their hike to Visp, and stop to rest under an overhanging stone cliff after passing through Mörel. Here, they scratch their initials and the date into the stone. On the veranda of the hotel in Visp, "*an unsuspected pleasure was awaiting us, namely the alpenglow, which was quite wonderful to see from so close up: at first, the distant snowfields were tinted purple, whilst the bare cliffs shimmered in the most splendid violet; then the golden light gradually died down; the glowing red changed into a pink colour, became paler and paler, and finally lay like a gleam on the uppermost tip of the* Jodlerhörnli". At breakfast, not a single cloud is to be seen in the summer sky. At the Visp station, the train to Zermatt is waiting, but that luxury would have cost the two brothers 20 Francs, so that they prefer to proceed on the soles of their boots to Zermatt.

Via St. Nikolaus, they arrive at Randa in the evening. On the fourth floor of the Hotel Weisshorn, they are given a pretty room with a view of the *Brunegghorn* and the *Weisshorn*. Their mood is excellent, the sun is just setting, and Zermatt is quite near. But what a disappointment—when they awaken, thick clouds have gathered, and not a single mountaintop is to be seen. They continue on over soggy roads to Täsch, where they have to seek shelter in a barn from the beginning rain.

Marcel and Eugen are taken aback by the poverty of life that they observe in many parts of the *Valais*. In the 1940's, Eugen, as a finance expert, has a part in the introduction of old-age and survivors' insurance; it is quite possible that the impressions of this early walking tour have awakened his social conscience.

The two brothers are now almost at the final goal of their hike: Zermatt lies just ahead. Nothing is to be seen of the Matterhorn, however; it has been swallowed up in the fog. In choosing a hotel, they have to be quite careful: "*In spite of the large number of hotels, making a choice was rather difficult; for we were told to pay close attention and use all of the experience that we had gained thus far, in order to avoid getting into an expensive barracks. [...] So we entered a hotel which had a simple aspect from the outside—but appearances were deceptive.*" Their room was granted no pardon: "*This monster had nothing like a well-defined shape; it was under the roof and it was impossible to stand up or even sit upright except in front of the door. The article of furniture 'chair' was represented by a single, poor-quality object, and beyond that, the bed clothes, towels etc. were thickly covered in soot.*" The boys are anything but happy over their domicile, but it has already begun to rain again.

On the following day, the dry moments are again rare, but the brothers explore Zermatt in spite of the dampness, and even get to see part of the Matterhorn for a few minutes before another cloud covers the mountain again. Since the weather remains wet and unfriendly, they decide to depart after a couple of days. Their mood is at a low point; their stay in Zermatt has cost time and money, and they have seen practically nothing. They go back to Visp, where they sit for a long time in the comfortably-heated hotel lobby reading newspapers. When they wake up the following day, their situation is—one can hardly believe it—"*by a few degrees still more difficult and unpleasant. It was raining, indeed—no—it was pouring in bucketfuls*". This poses serious questions about the planned route back over the *Gemmi* Pass; travelling by train via Lausanne and Bern to Basel would be the alternative. After breakfast, Eugen sends a telegram to Basel, asking permission for this change of route and for money to pay for the train tickets. They are in fact not quite so self-sufficient as they had been feeling. Their parents reply to their telegram around noon; they leave it open, either to wait for better weather in Visp or Leuk, or to travel back by train. The money will be sent by telegram. "*That was great to hear!*"; Eugen and Marcel are pleased and first off, they celebrate with a substantial meal for lunch.

The rain finally stops, and they decide to spend the night in Leuk. There, they enjoy, in view of the fact that "*the holiday trip is practically coming to an end,*" a 'last meal', i.e. a "*solonnel* [festive] *dinner.*" After breakfast the next morning, the weather has so far improved that they can in fact attempt the

Gemmi Pass, and through gorges, passing by alps and meadows, they cross over the pass and descend on the Bernese side near Schwarenbach into the *Kandertal*. "*The region was much less romantic than on the other side of the* Gemmi, *but also much friendlier.*" Striking farm houses and orchards greet them, a picture "*of stolid prosperity*". They spend the night in Bühbad and hike via Kandersteg through the lush green landscape. Unfortunately, they miss seeing the *Blausee*, although they pass quite close to it, which annoys them considerably in retrospect. It is reputed to be "*the greatest natural beauty in the* Kandertal, *and is said to be distinguished by its splendid play of colours and its picturesque surroundings, forested with fir trees and beeches.*" The way goes on to Waten Wimmis. There, in the "*Gasthof zum Niesen*", they obtain a room "*after repeated, in the end rather rough demands. [...] Only too soon did we discover what sort of people we were dealing with—the entire personnel was ... idiotic. Our room likewise looked highly idiotic. It was colossally large, almost a ballroom. The evening meal was—idiotic. There were long pauses, during which one heard the landlady and the cook emitting unarticulated sounds in the kitchen.*" After this damning criticism, the two stern wandering journeymen finally enter in dignified fashion upon their return trip from this not quite perfect holiday, and after several hours they reach Gurzelen, the final stop on their walking tour, during which they have covered about 250 km and 9000 m of altitude. "*After some enjoyable days there, we travelled by train via Bern-Olten to Basel* [23]."

There, father Jules has been examining several options for his future business ventures. He briefly enters the coal trade, and rents a "*splendidly-located plot of land*" from the Alsace-Lorraine Railway in Hüningen, where he sets up receiving and delivery of coal. But his bad luck continues: The following two winters are extremely mild, so that only small amounts of coal are required for heating. His income is pitiful, and his remaining assets are melting away. The recent coal dealer comes to the conclusion that "*this business is very nice and practically set up, but it has only one disadvantage, the main thing: It produces no profits.*" He plays for a while with the idea of a business near Paris; Edi is already looking forward to moving to the Seine, but it all comes to nothing. Finally, Marcel hears, presumably from a fellow student, that the cotton-wool factory Meyer & Co. in Thalwil, on the left bank of Lake Zurich, is up for sale. His father immediately meets with the family of the owners. They initially show little interest, but then in 1898, the transfer is sealed. Grossmann now puts all his eggs in one basket and, with his last financial resources and thanks to the help of Mimi's brother Hans Lichtenhahn and a friend, he acquires the small factory, which at that time employed six workers.

It is no coincidence that the factory is located in Thalwil: Cotton-wool is produced as a by-product of the cotton spinning works, and with the firms Heer, Robert Schwarzenbach, Dyeworks Weidmann, and the cotton spinning mill Schmid, Thalwil boasts of a whole series of textile works at the end of the 19th century. Cotton-wool has many uses, from medical bandages to shoulder pads for tailoring. The sector is subject to cyclic fluctuations, and Grossmann is well aware of the risks. However, since he—having learned from his experiences in Hungary—is the sole manager and at the beginning fulfils all the administrative functions by himself, he can work efficiently and relatively cost-effectively. In the coming decades, he will modernise the somewhat outdated enterprise with his characteristic energy and the vigorous support of his wife, optimising the production processes and improving the working and living conditions of his employees and their families. His predecessor had paid them poorly and as daily-wage workers, and there was a correspondingly high rate of fluctuation. Grossmann soon introduces a piecework system with long-term contracts, and everyone is happy with the results. He describes the factory as he took it over: "*In the downstairs carding room on the* Pilgerweg, *when I took over the business, the motor with its switching panel was to the right of the door, then two carding machines; to the left a beating machine, further to the left an old but good-quality vertical steam boiler. The department on the upper floor and towards the* Landstraße *consisted at that time of two parts; upstairs on the right were some stocks of cotton, and to the left were parts of the glue fabrication from old, greasy weaver birds, which were boiled and in an often evil-smelling manner used as glue for the finishing glaze. […] The raw material for the cotton-wool consisted of waste cotton from the spinning mills; this waste came and comes from the most diverse locations and facilities, and is thus extremely heterogeneous.*"

If Henriette had hoped that they could remain in Basel and Jules would commute from there to work in Thalwil, then she was disappointed. Directly next door to the factory building stands the roomy Villa *Rosenau* at the *alte Landstraße* 156. Jules argues that the family should move there, and so the two of them again "break camp" in Basel and move to the banks of Lake Zurich. Mimi is unhappy, but she agrees, and merely expresses her wish that they not have to move again. In fact, the couple will spend all the rest of their lives in Thalwil. Eugen accompanies his parents, Marcel gives up his room at Mrs. Hardmeyer's in Zurich and again lives with his parents, whilst Eduard remains in Basel with Pastor Hans Lichtenhahn until his confirmation.

As always, Mimi makes the best of the situation. She converts the Villa *Rosenau* into a pretty and cozy family residence. The stately three-storey house with a prominent gable is only four years old when Jules, Henriette and Eugen move in. It is surrounded by a terrain of 5000 m² of *"open, wild meadowland"*. Under Mimi's dextrous gardening hands, this prairie will develop into a romantic and practical garden, cut through by gravelled paths and planted with boxwood hedges, lilac bushes, roses, fruit trees and berry bushes, in the fashion of the times. A chicken yard is also established, and cats and the dog Caesar keep them company. The photo on page 58, taken on May 28th, 1899, shows the Grossmann brothers, Albert Einstein and Marcel's friend Gustav Geissler from his school days at the Gymnasium in Basel, in front of the Villa *Rosenau*.

In Thalwil, Jules and Mimi give proof of their flexibility and stamina. The two returned citizens from Hungary apply themselves without complaints and with little regard for their own health to their new work. For decades, the whole family gets up at five in the morning and has already eaten breakfast before six. Shortly thereafter, Marcel and Eugen leave the house in order to catch the train to Zurich. Their father works a twelve-hour day, and Mimi helps—along with the household and the garden—in the office; occasionally, she can also be found in the factory.

Although both of them work nearly around the clock, the situation remains precarious up to the First World War. They have no reserves, so every missed order can lead to serious problems. Jules feels responsible for his workers and pays them even when production is not used to capacity. This is not at all a matter of course: In the industrial countries, workers were at that time still to a large extent at the mercy of their bosses. They can be fired immediately and without compensation, if the boss so wishes. The business owners also have no financial safety nets, and Jules is well aware of the dangers. The spectre of "welfare" is always hanging as a threat, so the nearly sixty-year-old has to save urgently for retirement. At the same time, he wants to give his children a solid professional education. Jules' mother and sister are of the opinion that the younger branch should begin immediately to work and earn money in view of the meagre finances. An orderly apprenticeship as merchants or a solid craftsman's trade would be preferable to them to years of study; but the parents permit all their children to attend college. To accomplish that, they do without any kind of luxury, deny themselves an occasional theatre evening or concert, and holiday travel is not an option. At most, they permit themselves an occasional cure-weekend in

Weggis or Vevey, or a few days of vacation with Mimi's sister. Marcel and Eugen are allowed to study mathematics and law and economics; the latter can even spend a semester in Paris. Edi goes to a technical school and then gains practical experience in two technical enterprises in Münchenstein and Paris. Amélie's professional education is more erratic; following her breaking off the teachers seminar in Basel for health reasons, she is "in service" for a time with the family of a Prussian officer in Mecklenburg-Schwerin, presumably working as a governess and housemaid. After her return to Switzerland, she enters the school of arts and crafts in Zurich, but does not finish the course there.

Whilst Jules Grossmann is again devoting himself to his beloved motto *les affaires avant tout*, it is, as in Budapest, Henriette who is responsible for human relations. She forms acquaintances quickly and soon knows the whole neighbourhood. She thus also gets to know the Keller family, who live on the banks of the lake in their House "*Rebgarten*", and whose two girls will become her daughters-in-law. Marcel indeed discovered the older one, Anna, before his mother: The train between Zurich and Thalwil appears to have been a favourite location at that time for young people of opposite sex to observe each other from close up. The student Marcel speaks to Anna and quickly begins a flirt with her.

Eugen also has his eyes on a young beauty. She is the daughter of a silk producer from Rüschlikon and resembles the *Venus de Milo* in the eyes of the amorous student. Nearly all the young men on the train admire her and try to get a seat nearby. Only Marcel is the exception, and he reveals mercilessly to Eugen that the Venus of Rüschlikon is already spoken for. But Eugen continues to hope for a while until he finally turns away, disappointed. He doesn't have an easy time of it during the first years in Thalwil in other respects, also. German is indeed the mother tongue of the three sons, but owing to the origins of their teachers in Budapest, they speak a High German that is tinged with Saxon, with a pinch of Basel German and Alsatian. Eugen notes with annoyance that he and his brother are initially made fun of by their contemporaries in Thalwil because of their accents. Soon, they reluctantly accustom themselves to speaking the local dialect, even though for the rest of their lives they cannot speak the genuine Zurich dialect.

Photos Anna Keller (1893)|Anna and Hedwig Keller (1892)|Anna Keller's parents' house in Thalwil

Mimi Grossmann doesn't worry about such matters; she has quite different interests. Alongside her active work in the business, the living conditions of the workers and employees have a high priority for her, and she helps to improve them. She joins the Women's Club, supports destitute pregnant girls, and helps to establish a kindergarten. After a short time, she has an excellent network and feels at home in the beautiful country community. At the end of 1899, Eduard also moves to Thalwil, and for a short time, all three sons again live under the same roof, and Amélie also often visits.

Marcel and Eugen go through an intensive phase of rebellion against the beliefs of their parents. They discover philosophy, and the elder son delves into the scientific doctrines of Charles Darwin and Ernst Haeckel. He is fascinated by the theory of evolution and informs Eugen immediately of the results of these studies, which in Marcel's eyes can mean only one thing: The demolition of the biblical creation story and other dogmas of Christianity. Cheerfully, the two sons announce to their flabbergasted parents one Sunday morning that they find it superfluous to participate in the external customs of religion in future, and that they are therefore no longer willing to go to church. Jules and Henriette must have been completely dumbfounded. They have, to be sure, had some experience with difficult teenagers; Ernest has prepared them well in that respect, but even he never dared to present his parents with a *fait accompli*. There is a mighty quarrel in the Grossmann house; their father is beside himself and threatens to prevent both of them from attending college. "*I would rather have a devout shoemaker than a heathen scholar!*" he angrily shouts.

Mimi's brother, Pastor Hans Lichtenhahn, is called to their aid; a long correspondence begins between him and Marcel. Eugen recalls with a smile that his brother was usually the victor in these ideological disputes, and more than once, his uncle had no answer to the decisive logic of Marcel's argumentation. When even the grey eminence of the family can make no progress, the parents finally give up their efforts to save the souls of their sons. If they had secretly hoped that one of their offspring would take up a theological career, as had been thought to be quite possible in the case of Ernest, they are to be disappointed. In compensation, another, long-held wish of Father Jules is fulfilled: After he had tried in vain to interest Ernest, Marcel and Eugen in a career as a merchant at his side, and Amélie had demonstrated only mediocre talents and little enthusiasm, the youngest son, Edi, decides to follow in his father's footsteps. From 1906 on, he manages the business together with his father, even though Jules interferes in every important decision up to the end of his life, and Edi is permitted to act only as the junior partner for nearly thirty years.

5

Encounters

When Marcel begins his studies in October 1896 at the "Federal Polytechnic Institute", he is young and enterprising. The same could be said of his new college: It is just 41 years old, and thus a baby in comparison to institutions like the *Georgia Augusta* in Göttingen, founded in 1734, or even the University of Zürich, which began teaching in 1833, [not to mention the venerable universities of Heidelberg, Paris, Oxford or Bologna]. Furthermore, as its name implies, it is a technical college, and has no faculties with long traditions in Ancient Languages, Theology or the Humanities. Its focus lies in the education of a technical, practical intellectual elite—precisely what a modern society needs at the time of the [second] industrial revolution. Here, there is a young and fresh spirit which attracts teachers and students from all over Europe.

A retrospective glance at its beginnings shows that it was quickly implemented following the decision of the National Council, mentioned above. A year later, in 1855, the Polytechnic Institute was founded. Its funding comes from the Swiss Federal government, i.e. the Swiss state, in contrast to the universities, which are supported by the Cantons. Elwin Bruno Christoffel establishes the professional school for mathematics and physics. A dignified building is desired for public representation; Gottfried Semper wins the architectural competition for the prominent edifice, which sits enthroned above the city and is completed in 1864. It is not only impressive owing to its architectural allure, but is also very practical, with modern laboratories, collections and lecture halls.

© Springer International Publishing AG, part of Springer Nature 2018
C. Graf-Grossmann, *Marcel Grossmann*, Springer Biographies,
https://doi.org/10.1007/978-3-319-90077-3_5

Photo Marcel Grossmann, Albert Einstein, Gustav Geissler and Eugen Grossmann, *Rosenau*, Thalwil (from left to right, 1899)

Photos A postcard from Marcel Grossmann to Pauline Einstein (both sides; March 28th, 1899)

The Polytechnic Institute is one of the foremost and best-equipped technical colleges in the world. The domed building which dominates the cityscape today stems from a later period and was built in 1915–1924 by Gustav Gull. Where today, in the modern ETH, nearly 18,000 students and graduate students study at the two sites, Centre and Hönggerberg, in the winter semester of 1896/97 there were only 841 regular students [24]. This included co-eds: It was possible for women to study here from the beginning. The Polytechnic was only the second college in Europe to admit women, after the *Faculté des Lettres* in Lyon. Since however there were at that time no high schools for women in Switzerland, in the early years only women from abroad could take advantage of this liberal admissions policy.

The young student Grossmann matriculates in Section VI A, the "School for Specialised Teachers in Mathematical and Scientific Areas", where mathematics, physics and astronomy are taught. His class has gone down in history—not because of the modest number of students, only 11, but because along with Jakob Ehrat, Louis Kollros, Mileva Marić and Marcel Grossmann, a certain Albert Einstein is matriculated here. This small class size seems idyllic in comparison to the overfilled lecture halls of today. Instruction is correspondingly individual and personal. In the lecture halls and laboratories of the Polytechnic, close friendships between the students are formed.

The level of instruction is first-class, and the mathematics education under the professors Hermann Minkowski and Adolf Hurwitz can be considered to be among the best in all of Europe. Minkowski, a German, was born in 1864 in what is today Kaunas, Lithuania; he is 32 years old and recently appointed to his professorship at the Polytechnic. In Königsberg, at that time part of Prussia [now *Kaliningrad*, Russia], he had met Adolf Hurwitz, five years older, and become friends with him. The third member of their group is David Hilbert, who was born in Königsberg and became one of the most famous mathematicians of his time. Although Hilbert never taught in Zurich, Minkowski and Hurwitz are colleagues there for some years. Minkowski in particular makes a strong impression on the students because of his pedagogical skills and his brilliance.

Photo Anna and Marcel at the time of their engagement (1901)

Photo Marcel (around 1897) | August Friedrich Lichtenhahn with his son
Johann Friedrich and Marcel Grossmann, Braunau TG (around 1899)

Whilst Einstein organises his studies very independently and occupies himself with the latest developments in physics according to his own tastes (shortly before his death, he wrote: "*For people of my type, of brooding interests, studying at a university is not necessarily beneficial*"), Marcel is a classic model student, who absorbs in particular mathematical knowledge like a sponge and carefully takes notes in the lectures. His lecture notes are still kept at the ETH today and are in the meantime available online. They provide a unique insight into the mathematics courses at the Polytechnic around the turn of the last century. He generously allows Albert to use his notebooks before examinations when asked. The final examinations, in early 1900, are taken by Marcel, Albert, and five other fellow students out of the original eleven who started in the same semester. Grossmann finishes his exams as second best in the class. On the usual Swiss scale of 1 to 6 (best note), as already used on his certificate from Basel, he achieves an average of 5.23, whilst Jakob Ehrat takes third place with 5.14, and Albert barely makes fourth place with 4.91. Mileva, Albert's future wife, fails by a narrow margin and has to repeat the examinations. Louis Kollros, with an average of 5.45, is the best of the class; he is the son of a baker from La Chaux-de-Fonds, but his family comes from Prussia, a frequent circumstance in the former Prussian Principality of Neuenburg (Neuchâtel). From 1909 on, he will occupy the chair for descriptive geometry and synthetic geometry in French at the ETH. Kollros is later a great help to his colleague and close friend Marcel Grossmann, and he substitutes for him when Marcel's health begins to fail.

A new century has just begun, most of the class have obtained their diplomas, the semester holidays lie before them—and Zurich is a thoroughly pleasurable spot at the turn of the century. It boasts of no fewer than 1029 taverns, beer and wine bars, coffee houses and pastry shops. And they enjoy a brisk trade, to the chagrin of the authorities, who fear a decline of morals due to the low prices of alcoholic drinks and their correspondingly high consumption [25]. An exuberant mood prevails in those years before the First World War, in widespread regions of Switzerland: One festivity follows another, from children's parades to gymnastics meets to singing and music festivals and sharp-shooter competitions.

The students of the Polytechnic also enjoy life to the hilt. Once a week, they can be found in the fashionable Grand Café Metropol at *Fraumünsterstraße* 12. In the high-ceilinged great hall with its glass domes, gilded mirrors and monumental paintings, they for the most part spoon up iced coffee and feel like men of the world. During these light-hearted hours around the café tables or, in the unusual case that their finances permit, in the dining room with its heavy wooden chairs and colourful tablecloths, friendships that will last a lifetime grow between these young men. They converse on Life, the Universe, and all that, and jokes and scorn are never in short supply. The student Louis Kollros, who is frequently part of this scene, much later recalls that time with the words: "*I had the privilege of being Marcel's fellow student. He impressed me immediately with his liveliness and energy; he understood everything at an amazing pace; work was like a game for him; he kept us on our feet with his cheerfulness, his critical spirit and his trenchant manner of expression; nothing escaped him, neither the minor weaknesses nor the qualities of his fellow students, nor of our professors [26].*"

Already in the early years of the friendship between Einstein and Grossmann, it becomes clear that these two both love quick-witted and clever dialogues, and can play theories and ideas back and forth like ping-pong balls. The mathematician is fascinated all his life by the physicist and offers him much understanding over the years. His quickness of thought, his flashes of inspiration, the unconventional style of life and the inimitable sense of humour—all of that pleases Marcel, who himself is much more orderly, well-grounded and "reasonable". At least in later years; for shortly before the turn of the century, Grossmann was also a sly fellow, liked to play practical jokes and to use his charm on the young ladies. He grows a short, well-groomed beard, which suits him well.

His studies demand most of his energies, but he still finds time for many other activities: by day, he is a model student, but in the evenings, he dates a pretty waitress whom he met at the Grand Café Metropol and with whom he maintains—as Eugen records—a by no means merely Platonic relationship. At the same time, he is courting Anna Keller, and takes every opportunity to be close to her. How often did he run to the station from his last lecture so as to get a seat next to her on the train? Or walk past her house on a Sunday, in order to catch a glimpse of her? For a time, he seems to enjoy this unequal three-way relationship. "*Il faut que jeunesse se passe*", his father would have said, smiling suggestively—then Marcel makes his choice. The freshly-graduated mathematician doesn't climb up a mountain to make his decision, like his father before him. Spontaneously and with confidence, he asks Eduard Keller for the hand of his lovely daughter and evidently receives a happy assent. The young man must have been nearly euphoric. The 20th century is on its way, his professional future lies promisingly before him, and now he has a reason to celebrate: Marcel and Anna announce their engagement in 1899.

Albert Einstein is also in love; he and Mileva Marić have become intimate during their student years. He admires his intelligent fellow student from Serbia, is fascinated by her austere beauty, her soft voice, and her sometimes inscrutable manner. Albert writes her tender and passionate love letters and woos her in the most dedicated fashion.

6

Parallels

Soon after Marcel Grossmann obtained his diploma on the 27th of July, 1900, as a specialised subject teacher of mathematics, he initially took a position as assistant to Prof. Otto Wilhelm Fiedler, who had occupied the chair for descriptive geometry and synthetic geometry at the Polytechnic since 1867. This is a quite normal step for young graduates in the academic world. They can be of service to and with their professors, and at the same time work on their doctoral theses. The Polytechnic Institute received the right to grant doctoral degrees only on the 21st of September, 1908. Before that date, graduate students had to submit their doctoral dissertations to some other college or university. Whilst his fellow student Ehrat also obtains an assistant position at the Polytechnic, and Kollros accepts a teaching position at a Swiss high school before following Minkowski to Göttingen, Albert Einstein continues to seek similar employment, at first in a relaxed manner, then perplexed, and finally with increasing desperation. How alone he really feels at that time can be seen 36 years later: In his letter of condolence to Grossmann's widow Anna, Einstein recalls that he was on the verge of "*atrophying intellectually*", and his friend Grossmann had been a kind of "*lifesaver*" through his loyalty and his practical assistance.

Einstein has severe financial problems, and at first goes back to his parents' home. They are living in the meantime in Italy, at the *Via Bigli* 21 in Milan. Mileva remains in Zurich; she is worried about Albert's and her own future, but is at the same time determined to acquire her diploma, and will spend an additional year at the Polytechnic as a repeater. Albert is pursuing

© Springer International Publishing AG, part of Springer Nature 2018
C. Graf-Grossmann, *Marcel Grossmann*, Springer Biographies,
https://doi.org/10.1007/978-3-319-90077-3_6

his theories and ideas, writes often, plays the violin, and experiences alternating feelings of enthusiasm and doubts. In April of 1901, he writes to his friend in Zurich:

Dear Marcel! When I found your letter yesterday, I was deeply moved by your devotion and compassion, which did not let you forget your luckless old friend. I truly believe it quite unlikely that anyone could have had better colleagues than I had in you and Ehrat. [...] For the past three weeks, I have been here with my parents, in order to try from here to find a position as assistant at a university. I would have long since found one, had it not been for Weber's underhandedness. All the same, I leave no stone unturned, and do not give up my sense of humour [...] God created the donkey and gave him a thick hide. [...] As for science, I have had a couple of splendid ideas, which now only need proper incubation. I am now convinced that my theory of the attractive forces between atoms can also be extended to gases, and that the characteristic constants for nearly all the elements can be determined without too much difficulty. Then the question of the internal kinship between intermolecular forces and Newton's action-at-a-distance forces will be a great step closer to being solved. It is possible that experimental investigations already carried out by others for different purposes will suffice to test this theory. In that case, I shall utilise everything thus far known about molecular attraction in my doctoral dissertation. It is a glorious feeling to perceive the unity of a whole complex of phenomena, which appear to the senses as quite separate things.

I ask you to greet your worthy family members kindly from me, and thank your Papa heartily for his efforts and for the trust which he thus placed in me by writing a recommendation for me ... [27].

The Weber mentioned here was Einstein's favourite enemy figure at the Polytechnic, Prof. Heinrich Friedrich Weber. Einstein had initially greatly admired him, but made himself unpopular with Weber by jauntily addressing him as 'Mr. Weber' and frequently "forgetting" his professor's title. Whether Weber had actually taken that so seriously that he had consciously made Einstein's life difficult because of it can no longer be verified. The fact is that the physics student Einstein made a reputation for himself at the Polytechnic as a highly intelligent, brilliant thinker, but he could also be careless, bothersome, and recalcitrant.

In the winter of 1901 and 1902, Einstein publishes two papers in the "*Annalen der Physik*", a physics journal with a long tradition; later, he calls them "*my two worthless beginner's works*". He starts his first attempt at a doctoral dissertation, then a second. Both are unsuccessful; the angry author presumes that Weber is again behind this lack of success and writes to Mileva: "*What these Philistines lay in the path of someone who is not one of their own sort is truly dreadful.*" He postpones his dissertation for a while. But he can count on his friends: The "*efforts*" of Jules, which were mentioned by Albert in his letter, will finally bear fruit a little less than a year later.

Since he cannot obtain his doctorate at the Polytechnic, Marcel submits his dissertation in 1902 to the Philosophical Faculty II of the University of Zürich, with the title "*On the Metric Properties of Collinear Structures*". The work is supervised by Prof. Wilhelm Fiedler, the professor for geometry at the Polytechnic Institute. But since Fiedler does not teach at the University of Zürich, the task of being second reviewer of the thesis is assumed by Adolf Weiler, who is an adjunct professor there [28]. Following his oral examination, in June 1902, at the age of 24, Grossmann is granted the title of Doctor of Philosophy. From Frauenfeld, he receives an acceptance letter as a specialised teacher at the Thurgau Canton School. On the 2nd of December, 1901, he obtains permission to move to that Canton, as was required at the time of all those moving between different parts of the country.

UNTER DER OBERHOHEIT DER BEHÖRDEN UND DES VOLKES DES KANTONS

UND IM NAMEN DER

UNIVERSITÄT ZÜRICH

HAT DIE

MATHEMATISCH-NATURWISSENSCHAFTLICHE SEKTION

DER

PHILOSOPHISCHEN FAKULTÄT

IN IHRER SITZUNG VOM 3. JUNI 1902

DEM HERRN

MARCEL GROSSMANN

VON HÖNGG ZÜRICH

AUF GRUND SEINER DISSERTATION, BETITELT:

„ÜBER DIE METRISCHEN EIGENSCHAFTEN KOLLINEARER GEBILDE"

UND DER VORSCHRIFTSMÄSSIGEN PRÜFUNGSAUSWEISE

DIE RECHTE UND WÜRDEN

EINES

DOCTOR DER PHILOSOPHIE

VERLIEHEN

UND STELLT ZUM ZEUGNIS DESSEN DIESE MIT DEM UNIVERSITÄTSSTEMPEL VERSEHENE URKUNDE AUS

GEGEBEN IN ZÜRICH

24. JUNI 1902.

FÜR DEN AKADEMISCHEN SENAT   FÜR DIE II. SEKTION DER PHILOSOPHISCHEN FAKULTÄT

DER REKTOR: DER DEKAN:

Photocopy Marcel Grossmann's doctoral certificate (1902)

Einstein had also applied for the position in Frauenfeld, but in vain. His financial situation is becoming more and more precarious. Mileva has been pregnant since May, 1901, and Albert has no idea how he is to support the young woman, himself and the baby. He finally obtains two temporary positions, first as a substitute teacher in Winterthur, then as a private tutor in Schaffhausen. In the late summer of 1901, he writes hopefully to Marcel: *"But now I too am in the happy position of having gotten rid of the perpetual worry about my livelihood for at least one year. That is to say that as of the 15th of September, I will be employed as a tutor by a teacher of mathematics, a certain Dr. J. Nüesch, in Schaffhausen, where I'll have to prepare a young Englishman for the* matura *[his final examinations]* [29]*."*

The year becomes, in the end,—one semester. Einstein gets along famously with his pupil, but he falls out with the host family. In February 1902 he again departs Schaffhausen. In Bern, he had applied for a position as technical expert third class at the Swiss Federal Office for Intellectual Property, the Patent Office. He doesn't wait for an answer, but moves directly to the capital.

Einstein has high hopes, since he knows of Jules' *"efforts"*: The latter had recommended the young physicist to his friend of many years from Budapest days, Friedrich Haller, and asked him to consider Einstein for the next available position. After his return to Switzerland, Haller had become the first director of the Patent Office. With this recommendation and his diploma in his pocket, there is no obstacle to his being hired; Einstein has even acquired the necessary Swiss passport the previous year. His naturalisation as a Swiss citizen had succeeded after a police detective investigated him and reported that Einstein was *"an eager, diligent and extremely solid man"* [30].

Indeed, Albert receives the hoped-for acceptance to the position at the Patent Office. At last he has a steady income and some perspective! In the meantime, Mileva has given birth in January 1902 to a little girl in her homeland. Einstein never saw his daughter; she remained in Serbia, and her fate is still unknown today. It seems most likely that she died as a small child.

With Marcel Grossmann's help, the situation for Albert and Mileva begins to improve from 1902 on. Albert takes up his post at the Patent Office, begins to establish himself professionally and scientifically, and acquires a circle of good and inspiring friends, most especially Michele Besso, Maurice Solovine, and Conrad Habicht.

For Marcel as well, things are proceeding apace: He is working as a subject teacher in Frauenfeld and begins to publish scientific articles. He visits his fiancée as often as possible; she continues to live with her parents in Thalwil. On the occasion of the silver wedding anniversary of Jules and Henriette, he proudly escorts her to the extensive ceremonies. Both families are present in large numbers, and Hans Lichtenhahn, Mimi's brother, declaims a long and humorous poem which he has written himself in Baseler dialect, telling the story of the 25 years of married life of the couple with all its ups and downs. The young love of Marcel and Anna is also mentioned in a tongue-in-cheek fashion: "*Wenn gar vom Städtli Frauefeld der Professor und Briitigam sich stellt und luegt, ob am See no d'Liebi brennt, so waiss i nit, wo ner zerscht hi rennt, ob in Rebgarte-n-abe zur liebe Bruut oder zur Muetter ihrem guette Kruut.*" ["When even the professor and bridegroom turns up from the little town of Frauenfeld, and looks to see whether love is still burning on the banks of the lake, I still don't know where he first runs to, whether to the *Rebgarten* to his loving bride, or to his mother's good cooking."] Henriette's cooking skills are apparently unquestioned within the family and in her circle of friends; Eugen's enthusiastic memories of dinners at his parents' home in Budapest seem to be justified. She passed some of her recipes on to her daughters-in-law, for example the "*Pester Gipfeli*":

Recipe for *Pester Gipfeli*, according to Henriette Grossmann-Lichtenhahn:
1½ lbs. flour is mixed by hand with yeast for 10 lts [presumably litres] in a mixing bowl and then folded together with ¼ lb. butter, several tbs. sugar and a pinch of salt. 4 eggs and ca. ½ litre of milk are stirred in and the dough is allowed to rise in a warm place. Then it is rolled out and cut into squares or rectangles, which are spread with various kinds of jams and jellies (but only thinly), rolled up into conical shapes, brushed with egg yolk and garnished with chopped almonds and sugar. The *Gipfeli* are placed on a buttered baking sheet and allowed to rise for a short time in a warm place, then baked at moderate heat.

Photocopy Handwritten recipe for "*Pester Gipfeli*" by Henriette Grossmann (front side, frugally written on the back of an unused order form)

UST 1908

atten für Herrenkleider

ı Produzenten

-LICHTENHAHN

THALWIL bei Zürich

ɔnsumenten.

eibung:

Watten-Stärke

	Ko. 1.65 per 1 Stk. von ca. 0.90×12 m oder Ko. 1.00 per 1 Dtz. Blätter von 0.90×0.60 m
1.65	
	Ko. 2.5 per 1 Stk. von ca. 0.90×12 m oder Ko. 1.5 per 1 Dtz. Blätter von ca. 0.90×0.60 m
2.5	
	Ko. 3.5 per 1 Stk. von ca. 0.90×12 m oder Ko. 2.00 per 1 Dtz. Blätter von ca. 0.90×0.60 m

Photocopy Back side of the *"Pester Gipfeli"* recipe

In 1903, the situation in Bern and Frauenfeld is so stable that at last weddings can be celebrated: On January 6th, after Mileva has returned to Switzerland, the Einsteins marry, and on April 14th, the Grossmanns. Marcel and Anna are married by his uncle, Pastor Hans Lichtenhahn. The days are forgotten when Marcel and Eugen threatened to never again set foot in a church. The reception is held at the *Katharinenhof* (later the

Thalwilhof, today the Hotel *Sedartis*) in Thalwil, and the wedding banquet at the Hotel Belvoir in Rüschlikon. Both Mileva and Anna become pregnant soon after their marriages; the date of birth indicates that Anna was already expecting two weeks after her wedding!

On January 30th, 1904, Marcel Hans Grossmann junior is born, followed on May 14th by Hans Albert Einstein junior. Both babies bear the first names of their fathers and as second name 'Hans', a notable symmetry, which however may be coincidental. Although the two boys are of the same age and later go to school together at the *Realgymnasium* in the *Rämistraße*, they are not particularly good friends. Marcel junior notes in his memoirs: "*In my class was then also Albert Einstein, the older son of the famous Einstein (later Prof. for civil engineering in the USA)* [for hydraulic engineering, in fact]. *Since our fathers were great friends and worked together scientifically, they would have been happy to see us continuing their friendship. But that simply wasn't possible, we were too different* [31]." But at this point, high school is still far off, both children are still wearing nappies, and in a card sent from Bern in April, 1904, Albert writes:

Dear Marcel! Quite belatedly, I congratulate you with all my heart on the birth of your son, and thank you for having sent me your most recent paper, which I will certainly study when I have some time to devote to non-Euclidean geometry. Your solutions look very simple and elegant.

There seems to be a remarkable similarity between us. We are also expecting a baby in the coming month. And you will also receive a paper from me, which I sent a week ago to Wiedermann's Annalen. You treat geometry without the parallel axiom, and I treat the atomistic theory of heat without the kinetic hypothesis.

Did you know that Ehrat is teaching at a private school near Göttingen? He is quite happy there. And I am happy for him that he has found a pleasant existence and no longer has to serve as a 'satellite'.

Please give my love to your wife and your son (although we don't know each other yet), and accept my hearty greetings yourself. Yours sincerely, Albert and his Co-ed. [32].

In 1905, a year follows which for Einstein is unique. On March 17th, he submits a paper to the "*Annalen der Physik*" in which he uses the light-quantum hypothesis to explain the photoelectric effect, and thus makes a decisive contribution to the development of the quantum theory. This article is to earn him the Nobel Prize in 1922. A few weeks later, he finally submits his doctoral dissertation to the University of Zürich, as had Marcel three years earlier. In it, he discusses a new method for determining the sizes of molecules. He dedicates his thesis to his friend Marcel Grossmann [33]. Only ten days later, he sends another article to the *"Annalen"*, in which he successfully characterises the phenomenon of Brownian molecular motion. And then, in a *Eureka*! moment, he solves a problem over which he has brooded for years, with the help of a suggestion which will transform physics fundamentally. His solution later becomes part of the basic foundations of physics under the name of Special Relativity. Einstein's article bears the title '*Zur Elektrodynamik bewegter Körper*' ['On the Electrodynamics of Moving Bodies'], and he has written it within only six weeks.

Many books have been written on the theory of Special Relativity; they would fill kilometres of bookshelves. In the Internet, the search topic 'Special Relativity' turns up 17,000 links. The formula $E = mc^2$ is flaunted from T-shirts and coffee mugs. Distilled down to its essence and greatly simplified, one can say that it deals with the principle of how time can be considered to be a fourth dimension—alongside the three spatial dimensions (height, width, depth), and affects the properties of moving objects and fields. This can be illustrated by a picture which Albert Einstein himself often used: A train is passing through a station without stopping. The passengers in the moving train observe the people waiting on the platform. They have the impression that they are standing still as passengers and the platform is moving backwards past them. The people on the platform have the similar impression that *they* are standing still and the train is passing by them in the forward direction. And both are true!

Einstein's thought experiments are of course more complex; there, the people in the train and on the platform have clocks, rulers and sources of light pulses with which they can exchange light signals and measure space and time. But the principle of relativity can already be demonstrated by considering the relative—and yet correct—perceptions of those two groups of people. The distinction between relative and absolute perceptions can be visualised as follows [34]: On the breakfast table, the teapot is standing to the left of the cup. This is true from my viewpoint, but from the viewpoint of the observer sitting opposite me, it is false; for him, the teapot is standing to the right of the cup. Left and right are thus relative and depend on the positions of the observers. But the fact that the cup is filled to its brim is the same for both observers, they are in agreement; this statement is absolute.

The role of light can be explained in terms of an example in which a light beam is sent from the tip of the mast to the deck of a ship. To a sailor who is standing on the deck, the light beam is exactly as long as the height of the mast. But for a fisherman who is observing the same event from the shore, the light beam traverses the height of the mast plus the distance travelled by the ship between the emission of the light at the top of the mast and its arrival on the deck. Result: There is no absolute time; rather, all moving objects have their own relative times ['proper time']. The formula already mentioned [which follows from the principle of relativity and the constancy of the speed of light] specifies that two quantities originally defined by physicists as independent, the energy E and the mass m, are in fact equivalent to one another [35]: A change in the energy of a system produces a change in its mass.

In summary, our judgements about spatio-temporal relations depend in a very specific manner on the respective states of motion of the observers, according to Einstein's new understanding. This goes unnoticed in everyday life, because the new effects are large enough to measure only when the relative velocities of the physical systems involved approach the speed of light. With his work on the quantum theory, on Brownian motion and especially on Special Relativity, Einstein had changed the physics of the 20th century in a fundamental way.

Photos Marcel Grossmann (1906)|Anna and Marcel Hans junior (1906). *Herrenbergstraße*, Zurich

Whilst Albert is experiencing a year which will go down in the history of science as his 'year of wonders' ['*annus mirabilis*'], Marcel is progressing at a less dramatic rate. On April 7th, 1905, two days before Marcel's 27th birthday, the young Grossmann family moves from Frauenfeld to Basel, to *Gundeldingerstraße* 205 [36]. The mathematician is now teaching in Basel at his old high school. In addition, he is *Privatdozent* [lecturer] in the Mathematics-Physics Section of the Philosophical-Historical Faculty of the University of Basel, which he can manage well in terms of time [37]. When in 1906 Fiedler, his former professor for geometry and supervisor of his doctoral thesis, becomes ill, Marcel fills in and substitutes for Fiedler in the winter semester 1906/07. When it becomes clear in the following summer that Fiedler will have to give up his teaching duties permanently, the Institute begins to search for a successor. The favoured candidate is Martin Disteli; this 44-year-old mathematician teaches in Dresden, and he turns down the position with thanks. Thus, Marcel Grossmann is selected to be a professor on the 22nd of July, 1907, only 29 years old. His initial appointment is for three years, as is usual [38]. His course has been set: Marcel and Anna Grossmann return to Zurich at the end of September and move into a rented apartment on the edge of town, at *Voltastraße* 29, in the *Fluntern* District.

Only a few years earlier, as the group of students were celebrating their diplomas, their optimism for the new century was boundless. But politically, dark clouds are now gathering over Europe. After the German Empire had modernised its fleet beginning in 1900, it feels hard-pressed by the *Entente Cordiale* between France and England, and enters into an alliance with Austria. When Russia allies itself with France and England in the *Triple Entente*, the mood becomes explosive.

At the same time, social unrest is increasing, since the unfettered industrialisation of the previous decades has begun to show its ugly side—a working class has been formed, who often live and work under miserable financial, health and sanitary conditions. In many places, the environment is polluted by coal-burning for heat and power, and by dumping raw sewage and wastes into rivers and lakes. As so often, an external 'common enemy' gives governments an opportunity to distract their people from internal stresses and maladministration within their countries. What goes through Marcel Grossmann's head when he hears the sabre rattling which is becoming louder and louder? As the son of a French citizen, who has grown up in the Austro-Hungarian monarchy, there must be two hearts beating in his breast. Or is it three: He would be happy to support Switzerland and regrets that he has been declared unfit for military service due to a few centimetres lacking in his chest girth.

In 1908, Albert Einstein obtains his *Habilitation* [certification of quali-
fication as a teacher at the university level] from the University of Bern.
However, he has to schedule his courses early in the mornings before beginning his regular work in the Patent Office. His lectures take place twice a
week at 7:00 o'clock in the morning, and he has all of three (!) students
attending them, including his true friend and colleague Michele Besso. The
double burden of his work at the Patent Office and preparing his lectures
is gradually becoming too much for him, and Einstein asks Grossmann
about his opinion of a possible teaching position at Winterthur. Fortunately,
nothing comes of this, for shortly thereafter, he has finally arrived at his
goal, owing to a recommendation and mediation on the part of his friend
Heinrich Zangger: In 1909, he is named adjunct professor at the University
of Zürich, and takes up his new position on October 15th! Marcel is happy
to see his friend once again in his home town. The two of them often go
together and with students after classes to a café; their new favourite is called
the *Café Terrasse*, on the *Sonnenquai*, and there they hold discussions, talk
shop, and philosophise, whilst smoking without pause.

On March 22nd, 1909, the second child of Marcel and Anna, their
daughter Elsbeth, is born. The little family needs more space, and they rent
an apartment at *Herrenbergstraße* 1 in the *Oberstrass* District, within walking distance of the University and the Polytechnic Institute. That quarter,
in spite of its central location, is quiet and pleasant. But on certain nights,
the calm is interrupted: Marcel junior later recalls with enthusiasm that the
fire station of the *Oberstrass* fire company was right across the street, and it
was "*gruesomely exciting when the fire company's alarm horns sounded. The men
hurried to the station and pulled out the various wagons by hand, pushing them
off at a running pace.*" The Einstein family is also growing, with the birth on
July 28th, 1910 of their second son Eduard.

In that same year, Marcel Grossmann joins the Zurich '*Constaffel*', and
his brother Eugen follows suit in 1912. This society, which together with
the Craftsmen's Guilds is responsible for the Zurich *Sechseläuten* and helps
keep the traditions of the medieval guilds alive, was re-founded in 1899 as
an association. This new beginning followed difficult decades; the membership had decreased drastically. In 1868, the society sold its time-honoured
guild house on the *Limmatquai*, and ten years later, it was dissolved. But just
before the turn of the century, the *Constaffel* experienced a renaissance, and
following its re-founding as an association, its membership was increased
by admitting academics and military officers. This offered a chance for
young men like the Grossmann brothers, even though they did not have the
Privileges of the City of Zurich, nor a father nor grandfather who had been

a guild member before them. Thanks to the sponsorship of Marcel's fatherly friend and colleague, the *Privatdozent* [lecturer] for mechanical science and construction Rudolf Escher, they are admitted to the association. They are now thus Associates and can participate in the activities of the guilds, whose high point is the traditional '*Sechseläuten*' parade through the inner city of Zurich. In 1937, the *Constaffel* will buy back its guild house; but early in the century, its social events take place in the ornate new *Tonhallensaal* [philharmonic hall], with its *Trocadéro* towers. The two brothers remain Associates for the rest of their lives. Marcel junior marches in the *Sechseläuten* parade at a young age, costumed as a standard bearer; in 1926, he also joins the association.

The ETH professor moves smartly on the social stage, often accompanied by his spouse Anna. It is to his advantage that he can express himself fluently in several languages. He learnt English at the Higher Middle School, he speaks French well thanks to his father Jules, and he still has a passable knowledge of Hungarian. On trips abroad and in his contacts with professional colleagues, he develops ideas about how to intensify and improve mathematics instruction at the ETH and in Switzerland in general. He sets up summer school courses for high-school mathematics teachers, like those which are already given successfully in Germany. These courses are intended to increase contacts between the universities and colleges and the institutions which prepare the young people for higher education. He is also hoping that general interest for his subject can be aroused by the courses, and spares no efforts to make them as attractive as possible. The open, uninhibited spirit at the Polytechnic Institute makes these efforts all the easier.

Einstein, after having taught and carried out his research at the University of Zürich for two years, receives a tempting offer of a professorship at Prague, and decides to accept it. From April 1911 on, the Einsteins live on the Moldau, but they do not feel at home in this foreign environment, and rather soon, with Grossmann's support, Einstein is exploring the possibilities of a return to Zurich. Albert writes to Marcel on November 18th, 1911:

Dear friend Grossmann! Quite certainly, I am in principle inclined to accept a teaching position in theoretical physics at your Polytechnic. I am extremely happy at the prospect of returning to Zurich. This prospect induced me a few days ago to turn down the offer of a position at the University of Utrecht which had been directed to me.

That I definitely share your opinion, i.e. that the students of the VIIIth Division of the Polytechnic are offered too little, or too little that is up to date, in their senior semesters; this you already well know. I would be delighted to be able to help to fill this gap. As to the date when I might begin my tenure there, I would wish that it coincide with the beginning of a semester, since that would be best for my teaching duties; that could for example be next fall.

Best greetings from my family to yours, and to your brother Eugen, from your friend Einstein [39].

On December 10th, a second letter follows:

Dear Marcel! I must write to you once again about my appointment to a faculty position in Zurich. I mentioned in my previous letter that the thing is not urgent, since I would request that my date for taking up the position would be not before next fall, and I had already refused the offer from Utrecht. I regarded that matter as completely settled. But now today I received a laconic letter from H.A. Lorentz in Leiden, our greatest contemporary colleague, who is also a personal friend, containing a query as to how far along my negotiations with the Polytechnic Institute in Zurich have proceeded. He also asked for a rapid reply. This is all the more surprising since I received a detailed letter from him only yesterday, from which it emerged that he had completely reconciled himself to my refusal of the position in Utrecht.

There can hardly be any doubt that H.A. Lorentz will try to convince me to go to Utrecht all the same. If I am not at all officially committed in Zurich, then as you can well imagine, it will be rather difficult for me to refuse him again. I thus beg you to see to it prestissimo that the negotiations be initiated.

I sincerely excuse myself for pestering you in this fashion. But I am in such a delicate position that I don't know of any alternative.

With best regards, your old friend Einstein [40]."

Marcel Grossmann indeed springs into action *prestissimo*, and confers informally with the Rector of the Institute, Robert Gnehm. In its meeting protocol of December 1911 [41], the Faculty Council names Albert Einstein as successor to the "*unfilled position*" of their mutual teacher Hermann Minkowski, who had left the Polytechnic in 1902 and gone to Göttingen, where, seven years later, at only 44 years of age, he was to die of a burst appendix. His teaching duties in Zurich had been shared internally among the remaining faculty in the Section [42].

In July of 1912, Albert, Mileva and their two sons return from Prague to the city on the Limmat. Mileva had not felt comfortable in Prague, and is rather relieved to be back in Zurich. They rent a roomy apartment at *Hofstraße* 116. In fact, all the prerequisites for a future of many years of teaching and research in Zurich would now appear to be in place. As of October 1st, 1912, Albert is Full Professor for theoretical physics at the ETH, and Marcel is quite happy to be able to work side by side with him.

Their wives also get along well, although they never become close friends and continue to use the formal form of address with each other.

But Albert has just received an extremely attractive proposition from Berlin. He has been offered membership in the renowned Prussian Academy of Sciences. He in fact knows that Mileva and the children would prefer to remain in Zurich, but his marriage has not been going well for some time; the enamoured young students at the turn of the century have become a quarrelling, discontented couple. And Berlin is not only professionally tantalising for Albert; his divorced cousin Elsa Löwenthal lives there with her two daughters, and he feels quite attracted to her.

To Marcel's disappointment, and in spite of numerous attempts on the part of faculty and students to change his mind, Albert decides to go to Berlin, although he has already taken up his position in Zurich. In the spring of 1914, he moves to the city on the Spree and Havel. Mileva, Hans Albert and Eduard initially accompany him, but they will soon return to Zurich. Here, the common paths of the two friends come to an end; their tracks, which have run parallel for a number of years, now diverge. But before that happens, their friendship experiences a spectacular climax: They formulate together and publish together the outline of a theory of General Relativity.

7

A Scientific Cliffhanger

The story that began in 1912 and ended three years later with a spectacular breakthrough can hardly be surpassed for suspense, and it changed the lives of both of the two friends [43]. Louis Kollros described these occurrences later, in his epitaph for his colleague and fellow student: "*In 1913, he [Marcel Grossmann] was pulled into a scientific spiral that gave him a great deal of work and considerable pleasure. His friend and fellow student Albert Einstein had already arrived at the theory which is today known as special relativity, but without using the tools of higher mathematics. But when he dared to proceed to general relativity and the theory of gravitation, he found himself confronted with such great mathematical difficulties that one day, he consulted his friend. Marcel Grossmann was able to show him that the mathematical tools for the new physics had already been developed in 1869 by Christoffel in Zurich, the founder and first dean of the mathematics and physics institute of the Swiss Federal Technical Institute [ETH]*" [44].

Kollros would have known; he was the one who heard Albert's famous call for help and passed it on to posterity: "*Grossmann, you have to help me, otherwise I will go crazy!*" Even if things didn't happen in precisely this way, it is certain that Albert Einstein, in his search for a generalisation of his special theory of relativity, had gotten stuck and wasn't able to continue. As on previous occasions in his life, he found an energetic helper in Marcel Grossmann. This time, it was Grossmann's mathematical knowledge which offered the key to further progress. The "*scientific spiral*" is a fascinating chapter in the history of physics and mathematics, and it took place in two acts:

© Springer International Publishing AG, part of Springer Nature 2018
C. Graf-Grossmann, *Marcel Grossmann*, Springer Biographies,
https://doi.org/10.1007/978-3-319-90077-3_7

<u>Act 1</u>: Zurich, Summer 1912. Albert Einstein and his family have just returned from Prague to the city on the Limmat. Ever since he published the theory of special relativity seven years before, this topic—alongside numerous other physical problems—had given him no rest. In 1907, he published a treatise in the "*Jahrbuch der Radioaktivität und Electronik*" on the "Relativity Principle and the Conclusions drawn from it". This was followed by further descriptions of the theory, for example in 1910 a modernised version, translated into French by Edouard Guillaume, which was published in the "*Archives des sciences physiques et naturelles*". In his further thinking on the topic, Einstein aimed to extend the theory so that it would also include gravitation, the force of gravity.

What may sound harmless is in fact extremely complex. Einstein persued this goal for many years. The fundamental idea is that he now considers accelerated systems [special relativity deals only with "inertial frames", which are not accelerated], and the inertial forces which result are seen to be equivalent to gravitational forces [the famous "equivalence principle"]. In Prague, Einstein developed this equivalence principle into a prediction which could be tested by astronomical observations. In continuing to track this idea, he examined and rejected countless paths. When he arrived in Zurich in July 1912, he directed his famous, half-joking and half-serious cry for help to his friend Marcel. The latter was indeed rather busy; he spent the end of August with Anna at an international conference on mathematics in Cambridge, England, and gave a lecture in September at the annual meeting of the Swiss Society for Natural Science Research (SGNF) in Altdorf.

Photo Conference participants in Cambridge (1912), Grossmann marked in picture

But the appeal that Albert made to him did not go unheard. Even much later, in the 1950's, Einstein remembered that Marcel was all for it. He searched around in the tool kit of higher mathematics, which had developed quite rapidly in the preceding decades, hoping to find suitable instruments. They should permit the description and calculation of the motion of particles in a gravitational field, and the dependence of that field on the distribution of the matter in it. The hand-written notes referring to Grossmann in the original documents are very telling in this respect. It is interesting that Marcel Grossmann in fact taught descriptive geometry, and these researches were far outside his usual fields of study. But now the quality of his mathematical education showed itself, and his conscientious and thorough studies bore fruit. Grossmann had an excellent familiarity with the mathematical literature and suggested unconventional solutions which would later prove to be ground-breaking. He guided Einstein's view toward new possibilities of representing the theory mathematically and of carrying out calculations.

In those weeks and months of intensive cooperation after Einstein's arrival in Zurich, the two fit together the pieces of a puzzle that would later go down in history. The scientific assessment by Prof. Tilman Sauer in the Epilogue of this biography elucidates the details of that cooperation. Its results were initially published in early 1913 as the "Outline of a Generalised Theory of Relativity and of a Theory of Gravitation"[1] in the *"Zeitschrift für Mathematik und Physik"*. This paper contains a physical and a mathematical part; Albert was responsible for the physical part and Marcel for the mathematical part. In June of 1913, a preprint appears and the two friends begin immediately to introduce the project to their respective scientific circles; they are in fact practicing marketing. As the word "project" makes clear, they themselves are quite conscious of the fact that the theory is not yet defined in its final form, but they are ahead of everyone else—and closer to their goal than they can imagine. The final formulation lies within reach. Whilst it is uncontested that the physicist Einstein contributed the lion's share to this genial theory, which is still uncontradicted today, the mathematician Grossmann provided him with indispensable instruments. This was an inter-disciplinary, unbureaucratic and fruitful cooperation in the best sense.

On October 29, 1912, Einstein writes to the mathematical physicist Arnold Sommerfeld: *"I am now working exclusively on the gravitation problem, and believe that I can overcome all difficulties with the help of a mathematician friend of mine here. But one thing is certain, that I have never before in my life troubled myself over anything so much, and that I have gained enormous respect for mathematics, whose more subtle parts I had considered until now in my ignorance to be a pure luxury!"* [45].

---

[1] The English translation of this paper can be found in the "Collected Papers of Albert Einstein" (CPAE), einsteinpapers.press.princeton.edu, Vol. 4, English Translation Supplement, Document 13.

In March 1914, Einstein left Zurich and moved to Berlin. Shortly before his departure, the two friends author another joint publication, which in May 1914 again appears in the "*Zeitschrift für Mathematik und Physik*" [46]. They take up the topic of the "Outline" again, comment on it and correct it in some specific points, since in the meantime, they have critically scrutinised some of their earlier assumptions.

Act 2: Berlin, 1915. Albert Einstein is by now a member of the prestigious Prussian Academy of Sciences and professor at the Berlin University [at that time still the *Friedrich-Wilhelm-Universität zu Berlin*]. He has cast a broad net over the theory of gravitation, developed and further extended the theory of the "Outline", and applied it to the calculation of quantities of interest in astronomy. In the autumn of 1914, he writes a treatise on the Einstein-Grossmann theory and uses for the first time the formulation "*a general theory of relativity*", changing its name from that in the "Outline". He begins his treatise with a mention of Marcel: "*In the past years, I have, sometimes in cooperation with my friend Grossmann, worked out a generalisation of the theory of relativity*" [47]. In April 1915, he complains to Levi-Civita about the lack of interest from the scientific world: "*It is odd how few colleagues feel the intrinsic need for a* precise *theory of relativity*" [48]. In the course of the summer, that will change. After Einstein gives six lectures on general relativity theory in Göttingen in June 1915, the fire is ignited. Other famous scientists such as David Hilbert begin to work on the theory. Now it is really an academic race against time; still friendly, but not without a certain rivalry. Independently of each other, Einstein and Hilbert discover problems with the earlier versions of the theory and try to modify it and develop it further in the sense of the original far-reaching goals. Who will make the breakthrough first? It is no secret that Albert Einstein was faster; he completed the theory of general relativity in November 1915. He effectively put in place the last remaining piece of the puzzle; the picture is now perfect. The physicist writes immediately to the mathematician Hilbert.

He makes it clear that he has returned to concepts that he had already examined together with his friend: "*... the only possible generally-covariant equations, which have now proved to be the correct ones, were already considered three years ago with my friend Grossmann. We gave them up, but only with heavy hearts [...].*" At the end of November 1915, he writes again to Arnold

Sommerfeld: *"But in the past month, I have had one of the most exciting and arduous periods of my life; to be sure, also the most successful."*

The breakthrough is successful beyond any doubt. Albert Einstein catapulted himself with this theory into a scientific league that knows no equal. After the heady and intensive cooperation in Zurich, the contact between the two friends practically ceases between 1914 and 1919, presumably also because of the outbreak of the First World War. A last indication of their correspondence is Einstein's letter from Berlin to his colleague Paul Ehrenfest in April 1914: *"Grossmann wrote to me that he is now also succeeding in deriving the gravitational equations from the general theory of covariance. This would be a nice addition to our investigation"* [49].

There is no indication that Albert informed his friend Marcel of the success of their cooperation. Einstein indeed expresses his thanks in a detailed exposé which he writes several months after the publication of the theory. But in the decisive paper itself, there is not a hint of Marcel's contribution. [In the first of the four papers from November, 1915, Einstein notes that he and his friend Grossmann had nearly arrived at the correct solution two years earlier; and he thanks Grossmann explicitly in the first summary article in 1916.] In a letter on July 15th, 1915 to Arnold Summerfeld, Einstein in fact bluntly plays down the cooperation with his friend: *"Grossmann will never make a claim to being a co-discoverer. He helped me with the orientation within the mathematical literature, but he made no material contribution to the results"*.

It would appear that Marcel shares this opinion; at least, he never emphasised his role in the theory. It is always Albert's achievements which he places in the foreground. Is this modesty, or does he see his contribution, as exciting as it may have been, simply as a temporary excursion into the field of physics, before he turns back to his core topics, mathematics and geometry? Is he truly not conscious of the fact that he fulfilled a real function as a bridge-builder between the worlds of the mathematician and the physicist? Most probably, he assessed the success of the theory with gratification and pride, but was already following other ideas and was involved in other projects.

Photos (2 pages) Group picture at the VIIIth Meeting of the Swiss Mathematical Society, Zurich, September 1917, 1: C. Carathéodory, 2: M. Grossmann, 3: D. Hilbert, 4: C. F. Geiser, 5: H. Weyl, 6: L. Kollros,7: F. Gonseth, 8: A. Speiser, 9: M. Plancherel, 10: H. Weyl

Einstein, in his thinking, is also already light years ahead. He corresponds with other colleagues, in particular with Paul Ehrenfest and his Dutch colleagues. Only in 1919 do the two friends again meet in Zurich, when Albert spends some weeks on the Limmat for a lecture series; and from 1920 on, they again take up their correspondence, and continue their familiar dialogue in letters and visits until the end of Marcel's life.

Their cooperative work in the fall and winter of 1912/13 remains remarkably fascinating for the ETH professor, and without a doubt was a high point of his life. But it also left him tired, in particular since he was busy with many other tasks at the same time. As early as 1913, Anna is the first to notice that Marcel sometimes drags one leg slightly when he walks. But her husband will hear nothing of resting; he feels well, and is only a bit tired. In the next vacation period he will again go on long hikes with his brothers, and will quickly recover. His father Jules would have remarked at this point, "*Les affaires avant tout*", even though in Marcel's case, one would instead have to say, "*la science avant tout*".

8

Dialogues

Marcel Grossmann is a teacher and professor, body and soul. His colleagues describe him as a talented, convincing pedagogue and docent. He never allows himself to lecture for hours on end without making sure that his students are understanding what he is trying to teach them—as *his* teachers did when he was in secondary school in Budapest. From his own school experiences, he knows that the pleasure of learning and the resulting success are incomparably greater when the material is taught in a gripping and readily understandable manner. In a report on geometry instruction at the Upper Middle School in Basel, written in the summer of 1898 during his studies at the Polytechnic, he maintains, "*In our middle schools, we have [...] teachers of mathematics, under whose instruction even the most beautiful subject would wither and dry up, & under whom one of the most elegant aspects of mathematics, its exactness in all directions, is distorted into pedantry. 90% of the pupils acquire a feeling of revulsion towards this 'dry' mathematics [...]. That one can awaken pleasure and interest in mathematical science quite early [...] even in average students, not to mention the more gifted ones, has been my own experience repeatedly with pupils to whom I have given private tutoring or preparatory lessons for an entrance examination. One must simply know how to motivate them*" [50].

His beginnings as a young mathematics teacher in Frauenfeld are promising, and at his former secondary school in Basel, he will have made every effort "*to motivate*" his pupils. In the inspiring surroundings of the Polytechnic Institute, his true talents can then fully unfold. As Louis Kollros noted, "*The first impression that he made on the students was that of a leader and coach. His courses were structured with a splendid clarity, and he knew how*

C. Graf-Grossmann, *Marcel Grossmann*, Springer Biographies,
https://doi.org/10.1007/978-3-319-90077-3_8

to generate enthusiasm in his listeners. [...] Grossmann made a serious effort to find interesting examples and exercises and to suggest elegant applications of the theory to his students. [...] Even after they had left the ETH, the students still had contacts to Grossmann, and he strenuously supported their efforts to find suitable employment. Everyone wanted to be his assistant!" [51]. Grossmann takes the students seriously (most of them were young men at that time) and carries on discussions with them at eye level. Not only in the lecture and seminar rooms, but also in private, over a cup of coffee or a glass of wine.

In the summer of 1911, the elegant *Grand-Café Odeon* opens on the *Sonnenquai* (now called the *Limmatquai*), and it quickly acquires a reputation as a favourite meeting place for academics, artists, writers and the intelligentsia from Switzerland and from abroad. The fact that the toilets were forgotten during its planning and had to be hastily added before its opening is of little interest to its customers. And even the derisive term *'Café Headcheese'* which is soon applied to the premises by the general public tends only to increase interest in it and to set it off from the plebeians; after all, what do *they* know of the *Art Nouveau* style or of the reddish-brown marble panelling which gave it its nickname?

The café has a cosmopolitan atmosphere, with its high ceilings, large windows, glittering crystal chandeliers and polished brass ornaments [52]. At that time, it occupied the full width of the building. In the basement, pastries are prepared in its own bakery, and there is a billiard room on the upstairs floor.

What are the topics of conversation at those round marble tables? Life, the Universe, and all that, of course; philosophy and art, politics and university gossip, physics and mathematics—and "descriptive geometry and projective geometry", Grossmann's areas of specialisation. In the *Odeon*, there are plenty of opportunities for object lessons. Geometry is descriptive when problems of representation are solved using the methods of construction and drafting; that is, simply 'descriptively'. Pencils, compass and triangles, along with India ink and water colours, have remained the instruments that are used in its instruction, right up to the present day. The goal is to analyse a research question and to find a conclusive answer. In this process, three-dimensional objects such as cubes, pyramids or spheres are represented two-dimensionally, in perspective, on paper or on a computer screen. A typical exercise proposed in a course might be [53]: *"A child's building block is given as a layout sketch at a scale of 1:2. Show, corresponding to the coordinate axes of the draft, the axonometric projection of the block at actual size."* Or else the students might be asked to show the mirror image of a light beam: *"The beam impinges at the point P on the mirror plane ε which passes through g.*

Construct the beam s which is reflected from ε." A relatively simple problem which Grossmann formulated for his students around 1914 states: "*Draw a sphere of given radius whose centre point lies in the basal plane. Show the north pole and the equator of the sphere*" [54].

A knowledge of descriptive geometry is essential wherever planning, construction and fabrication are to be carried out, from high-rise construction to machine design to the manufacture of all sorts of products. This discipline presumes that its practitioners have analytic and spatial-visualisation abilities. The professor is visibly pleased to be able to make his lectures lively and vivid, instead of dry and boring. He chooses examples that are challenging, aesthetically convincing and modern. When the ETH main building is being enlarged in 1915, Grossmann's students draw the domed roof designed by architect Gustav Gull. They practice perspective drawing by drafting views of villas in turn-of-the-century style, or of a drainage gallery. The young men draw defence facilities—very topical during the First World War—and reproduce the surroundings of the Linthal power plant in the Glarnerland. Looking at the carefully prepared and beautifully coloured plans and drawings, one can see that they were not the product of dry school matter, but rather that the view of the students had been sensitised to the harmony of high-rise and underground construction. Although in 1915, outdoor photography with 35 mm cameras is still in its infancy, Marcel Grossmann is already using photos for projects in the young discipline of photogrammetry, i.e. surveying by means of photos. All in all, the Cartesian elegance of mathematics and descriptive geometry, with which complex problems can be concisely formulated and solved, seems to have been very satisfying to him. Like mathematics in general, descriptive geometry forces its practitioners toward solution-oriented thinking. And a convincing representation usually anticipates the solution. Evidently, Marcel has in the meantime also improved his own skills in technical drawing; we recall his marks in that subject at school…

In 1911, Marcel Grossmann, 33 years young, is chosen by the faculty of Mathematics and Physics for Specialised Teachers to be their chairman. He has an excellent reputation, is said to be a born organiser and a man of action, is full of ideas, and always contributes to their realisation. He must have been flattered by the assignment. Some years later, as his colleague Louis Kollros recalls in his epitaph for Grossmann, the professors of the Department of Mechanical Engineering elect him unanimously to be *their* chairman. This is quite an honour, especially since luminaries in the field were also available for election! Perhaps his success was due precisely to this fact: He was not one of the members of that department, is not in competition with its professors, and is therefore more objective. One thing is certain: The son of a factory owner, who had admired countless plans, sketches and constructions even as a child, can readily identify with the challenges facing his colleagues in mechanical engineering, and he will always have a special affinity for that department.

In his own area of specialisation, during his twenty years as professor for descriptive geometry (1907–1927), Marcel Grossmann will supervise countless students; he refers to *"several thousand"* [55]. Descriptive geometry is required of all the students of the natural sciences and engineering, whilst synthetic geometry is a mandatory subject only for the prospective teachers. Small classes like those that Marcel and Albert had experienced have become rarer, but the social interactions between students and faculty at the ETH remain personal and direct.

Marcel also strives for dialogue in a quite different area: Up until 1910, there was indeed an association of the mathematics teachers in Swiss secondary schools, which actively pursued improvements in the teaching of mathematics, but there was no professional society of Swiss mathematicians with a scientific, university-level focus. The Swiss mathematicians thus had little influence in international scientific organisations. The chemists and physicists had gotten ahead of them: The Swiss Society of Natural Science Researchers (SNG) already had sections for the chemical and physical societies. Now, the mathematicians catch up quickly. In 1910, Marcel Grossmann, Rudolf Fueter and Henri Fehr join together to take the initiative and, having secured the support of other professional mathematicians, they call for the founding of a Swiss mathematical society and their own permanent section within the SNG. They have two goals from the beginning: An association of the Swiss mathematicians is desirable to prevent their being disregarded by international academic organisations; and the new society proposes to make a practical contribution by editing and publishing the collected works of the Basel mathematician Leonhard Euler, in commemoration of his 200th birthday. No sooner said than done: The appeal of the three men calls up a strong echo, and all together 82 membership applications are received. On the 4th of September 1910, the constituting meeting is held in the *Bernoullianum* in Basel. Marcel is president from 1915 to 1917, and is named Member of Honour in 1935 on the 25th anniversary of the founding of the Society, together with Fueter and Fehr, *"accompanied by minute-long applause"*. He cannot accept the honour in person, since he is by then *"chained to his sick bed"* [56].

The Swiss Mathematical Society, in the over 100 years of its existence, has today evolved into a modern scientific thinktank and an internationally networked organisation. A regular rotation of the members of its board of directors among the faculties of all the mathematical sections and departments in Switzerland, from Geneva to Basel, guarantees that all the departments and every part of the country will be represented in its presidency over the course of time.

Establishing the Swiss Mathematical Society is well underway at the beginning of the year 1914. Marcel Grossmann is on the faculty of the ETH, Albert Einstein is living and working in Berlin, and his wife Mileva and their two sons, who had followed him rather unwillingly there, will return in the summer to the city of Zwingli. Eugen Grossmann, Marcel's brother, who had been educated in jurisprudence in Zurich and Paris, and then spent some years in Bern, working there first as Canton Statistician and then as Director of the Office of Statistics, has also returned to Zurich. He has been named professor for statistics and finance at the University of Zürich, where, years later in 1944 and 1945, he will hold the office of Rector [57]. Eugen becomes Marcel's fellow campaigner in many areas and his ally in everyday, political and ideological matters. And precisely in this field, there is unfortunately no lack of topics for discussion.

Georg Kreis describes the situation in the summer of 1914 in his book "*Insel der unsicheren Geborgenheit*" ("The Island of Uncertain Security") as follows:

> On June 28th, 1914, the first shots were fired in Sarajevo, of which it was said shortly thereafter that they had set off the great European War. To be sure, the idea that there must be a war had already gained a contour and a basis, and it was felt that this war would even have good aspects, because it would – finally – put an end to the smouldering state of mutual distrust, and would lead to a clarification and purification of the relationships between the competing powers. [58]

In Switzerland, a strange mood has taken over the country: The people are both gloomy and hopeful at the same time, because they almost long for the open conflict to put an end to the unbearable tensions among the Great Powers. The opinion is widespread here that the war is necessary— and will soon be over. After the beginning of August, events occur rapidly: On August 3rd, the Federal Council orders a mobilisation of the Swiss army, and the next day, Switzerland declares its neutrality. Another day later, it can be seen in Belgium, likewise a neutral country, how quickly the tables can turn even for presumably safe countries: German troops enter Belgium on their way to France. On August 25th, 1914, the German army destroys and plunders the Flemish city of Leuven. 200 of its citizens are killed and the University, including the valuable contents of its library, is consumed by flames. Belgium's neutral status is however somewhat different from that of Switzerland, as Jörg Friedrich describes in his book "*14/18, der Weg nach Versailles*" ("14/18, the Path to Versailles"): "*Belgium was not a neutral, but rather a neutralised country; that is a different matter. A neutral country avoids hostilities of its own accord, it wishes to stay out of the conflict. A neutralised country has had its neutrality imposed upon it by interested powers*" [59]. But for Switzerland, the fate of Belgium must have been a shock.

In the following months and years, Switzerland can indeed defend its neutrality, but the multilingual country has fallen into a deep, internal conflict of loyalties: German-speaking Switzerland and the majority of the Swiss Federal Council show open sympathy for the goals of Germany and Austria, whilst Western Switzerland exhibits more solidarity with the *Entente* member France. In Zurich, at the beginning of the 20th century, every fifth inhabitant is German, and these apparently well-integrated people are inclined to patriotism. As born Swiss Abroad citizens with a father who openly shows his sympathy for France, Marcel and Eugen Grossmann experience at first hand how the opposing factions collide with each other, and how rifts open up within Switzerland, between the different parts of the country.

Already since February, Marcel Grossmann, together with his cousin Paul Lichtenhahn, has been a member of a society which makes just this question into its central concern: The two young men are among the 157 founding members of the *Neue Helvetische Gesellschaft* [the New Swiss Society, NHG]. The NHG emerged from the *Helvetische Gesellschaft*, an organisation founded in 1761. The founders at that time felt that Switzerland, "*this country which is developing well economically* per se, *is in need of a renewal in the political and intellectual sense*" [60]. They imagine a *Helvetism*, whose thinking goes beyond the language and regional boundaries, through which progressive and conservative goals can be brought into harmony. After 1848, their original goal of a unified, modern Switzerland seemed to have been achieved; the meetings of the society were discontinued. But a few decades later, it had become evident that this goal was more topical than ever.

The NHG was founded in 1914, on the initiative of several young people from Western Switzerland, including Gonzague de Reynold, Robert de Traz and Alexis François [61]. The basic principles given in the founding invocation were "*...considering the dangers which are threatening our national life, but full of confidence in the future of our country [...] to establish closer bonds among ourselves.*" The NHG promotes "*a national and open education, the sharpening of the public conscience in the battle against our infiltration by* *an exclusively materialistic attitude, and the maintenance of closer relations between members of different parts of our country as well as to Swiss citizens abroad, among themselves and with their homeland."* These principles are publicised in all the important Swiss daily newspapers and call up a widespread echo. Opinions are divided; whilst the liberal-conservative media in Western Switzerland and the *"Schweizer Bauer"* (the "Swiss Farmer") react in a predominantly positive manner, more free-thinking media such as the *"Genevois"* and the *"Neue Zürcher Zeitung"* (NZZ) regard the new society and its youthful idealism more sceptically. The socialist press completely rejects the founding principles of the society [62]. But the goals of the NHG are not too different from those of the country's government; on October 1st, 1914, the Federal Council publishes an *"Appeal to the Swiss Population"*, addressed in particular to the press, *"to practise a wise moderation"*, and warning against the *"weakening of the feelings of solidarity through an incautious, hostile and hurtful emphasis on that which separates us"* [63].

In December 1914, the poet and author Carl Spitteler gives a widely-acclaimed address to the NHG in Zurich, with the title *"Unser Schweizer Standpunkt"* (Our Swiss Standpoint). Later, it is held to be one of the most famous domestic political speeches in the country's history. In his speech, Spitteler prompts his audience (the manuscript suggests that there were both men and women present) urgently to promote a brotherly solidarity among the citizens of Switzerland, even though the sympathies of the German Swiss tend towards their German neighbours, whilst those of Western Switzerland favour the French and the *Entente* more strongly [64].

In the NHG, Marcel Grossmann and Paul Lichtenhahn find an ideal field of action for their patriotic feelings. The ingredients are indeed attractive: Young, committed men from all the regions of the country are hatching patriotic plans and are eager to prove to the generation of their fathers that their thinking is modern and responsible. The age limit for members is 40, which is however mostly a clever move on the part of Gonzague de Reynold designed to eliminate an undesirable candidate. But whatever one may think of arbitrary age limits, the spirit of the organisation is young, dynamic, and enthusiastic.

Many of its members are academics, often from upper middle-class backgrounds, who count as members of the Swiss elite and who want to advocate maintaining the unity of Switzerland. The members correctly recognise that this small country, with its four languages, which has been a federal state for only fifty years, can hold its ground amidst the great powers and their struggles for authority only if it remains unified. This does not imply that all the regious should become similar to one another; on the contrary:

It is an explicit goal of the NHG that the unique features of each region should be preserved [65]. Marcel Grossmann also sees a possibility here of taking action for his country, even if he is not admitted to the military. Paul Lichtenhahn and Marcel Grossmann are both involved in the central executive board of the society from 1916 on, and Marcel actively supports an internal Swiss dialogue among the students at the ETH.

In 1920, the NHG supports the membership of Switzerland in the League of Nations, that federation founded by the victorious Allies after the First World War. But in the course of the 1920's and 30's, the ideological contradictions within the NHG grow. Again and again, opposing opinions collide at a fundamental level, and values such as patriotism and nationalism seem to contradict democracy and freedom of opinion. The founding member Gonzague de Reynold proves to be an ambivalent personality. This young man, "*a count by birth and an author by inclination*" [66], was born in 1880 as Frédéric Gonzague, *Comte Reynold de Cressier* in Fribourg. The historian and publicist exhibits a charismatic and winning personality and a contagious enthusiasm. In the course of his long career, he campaigns passionately for his political views on Switzerland, which become more and more nationalistic and right-wing-conservative over time.

De Reynold distrusts democracy and fears that it will open the way to a subversive communist attitude in the population. Jean-Rodolphe von Salis describes his history teacher at the University of Bern in his "*Notizen eines Müßiggängers*" [Notes from an Idler] as a conservative, royalistic Catholic with a well-developed sensibility for poetry, who fought in vain against the 20th century [67]. This agile, rhetorically talented and sharp-tongued intellectual becomes the Vice President of the Commission for Intellectual Cooperation in the League of Nations in 1932, but in the course of the Second World War, he develops sympathies for right-wing, authoritarian forms of government such as was practised in Fascist Italy. Marcel Grossmann is no longer affected by this ideological trench warfare within the NHG; his interest for the society has cooled by 1920, especially since he has by then passed the 40-year age limit. The society itself—since 2007, it has been renamed the "*Neue Helvetische Gesellschaft—Treffpunkt Schweiz*" (New Helvetic Society—Rendezvous Switzerland)—has long since distanced itself from the right-wing sympathies of de Reynold and today supports the democracy and the unity of Switzerland, by furthering mutual understanding, formulating goals which support the national identity, and contributing to answering important questions which arise in domestic issues or in the foreign relations of Switzerland [68].

When the First World War is finally officially ended in Versailles on June 28th, 1919, precisely five years after the assassination in Sarajevo, the conflict has brought immense suffering onto the world, caused the breakdown of the political, societal and financial structures in Europe, and has cost the lives of 15 million people. But the universities have continued to function, although under difficult conditions. The teachers and students gradually return and again take up their studies. It must have been rather difficult to find the way back to everyday activities after having spent four years on the front. Furthermore, even in academic circles, differences of opinion, conflicts of loyalty, and hidden or open racism put strains on the working atmosphere between colleagues. To add to the misery, the Spanish influenza marches across Asia and America to Europe. All together, it is estimated to have ended the lives of 25–50 million people. In Switzerland, 2 million out of the population of 3.5 million at the time become ill, and 25,000 men, women and children die of the disease. In these dark times, the dome on the extension of the main building of the ETH, which is in the meantime finished to the stage of a shell, acts as a ray of light and a hope for a better future.

Marcel Grossmann has by now been a professor for 12 years, and he has met students from all over Switzerland and from abroad. As had other faculty members, he has noticed the very diverse standards of education among the various students. How should he plan his instruction when a student from Chur has a very different degree of preparedness from that of another student from Basel? Since each Canton in Switzerland is responsible for its own educational system, this means that there are over twenty different yardsticks for the school leaving certificates! Up to now, the universities and the ETH have regulated access to university-level studies by contracts with the high schools, higher middle schools, and the industrial and trade schools. This complicated system is in need of simplification, and following the First World War, Marcel Grossmann, with great zeal, becomes a member of the Federal Commission which is charged with planning an educational reform. In his essay, "*Sinn und Tragweite der eidgenössischen Maturitätsreform*" [The Purpose and Scope of the Federal School-Leaving Certificate Reform], he describes the situation at the outset of the reform planning as follows, "*The structure of our federal state makes it inevitable that the reform of the middle schools is essentially a matter for the Cantons; and they, corresponding to the great diversity of regional circumstances, are faced with a wide variety of problems. The Federal government, in contrast, can exert an indirect influence through its guidelines for the school leaving certificates, by encouraging and facilitating a reform of the middle schools in an up-to-date manner.*"

In the autumn of 1919, the Commission instituted by Federal Councillor Gustave Ador begins its work, and about two years later it presents the draft version for a new law regulating the standards for school leaving certificates. This version envisages three types of school leaving certificates, which are described by Grossmann as follows: "*Type A of school-leaving certificates indicates that the student has matured through intensive studies of the two ancient languages and their associated cultural heritages; type B replaces ancient Greek by a modern foreign language, as chosen by the student; type C is characterised by more intensive studies in the field of mathematical and scientific subjects.*" Although the Commission has made a visible effort to formulate its recommendations in a balanced manner and to emphasise the autonomy of the Cantons, its draft is met by a concentrated charge of resistance. In particular, the importance of the ancient languages Greek and Latin is the subject of much discussion and leads to venomous disputes; some opinion leaders fear nothing less than the intellectual decline of the whole country. Pedagogical objections are expressed loudly; the suspicion is even voiced that critics of Catholicism are behind the relativisation of the study of ancient languages. Marcel Grossmann cannot resist taking a few forceful stabs in his essay: "*A simple woman of the people, who has raised and supported herself and her children through toil and trouble, certainly has a greater nobility than many a scholarly man who is ostensibly or presumably serving 'higher knowledge', but in fact has pursued only petty and self-serving goals throughout his whole life. And how many 'classically educated' men are there who have not felt even a hint of the free spirit of the ancient peoples, but rather persist as stodgy Philistines with narrow horizons amid their prejudices!*" Serious objections also come from medical circles; here, there is a fear that the new law will cause a "*flooding of the medical profession with unsuitable elements*" and will give rise to a "*foreign infiltration*".

The members of the Commission stand actively behind their draft version and publish articles like that by Marcel Grossmann just quoted. They however have to recognise with some disillusionment how difficult it is to introduce reforms speedily within a decentralised, democratic system. Even when there is general agreement over the analysis of a problem and the suggested remedies, every complex reorientation is ineffective unless it is put into effect in a unified manner. Only in 1925, after years of consultation, does the Federal Council enact a country-wide, general law governing the standards of school leaving certificates of the types A, B, and C; it then remains in effect for around five decades. The work of the Commission was thus successful in the end, but immediately after the completion of its work, many of its members, among them Grossmann, were frustrated. The ETH professor makes not only friends with his uncompromising, sometimes undiplomatic engagement, and the dialogue at times turns into a vehement exchange.

9

Anna

Marcel Grossmann engages himself on many fronts; he gives lectures and makes speeches, writes articles and textbooks, and purposefully establishes his social and professional networks. And, as in the household of his parents, it is the duty of the wife to organise all the other aspects of life. Anna Grossmann does this gladly; she finds the role of a professor's wife and her life at the side of her young, generally well-liked and good-looking mathematician husband pleasing. And she had been prepared perfectly for her duties.

The professor's wife was born in 1882 on the banks of Lake Zurich. With her two older brothers Max and Otto, and her sister Hedwig, 11 years younger, she grows up in the stately house at the "*Rebgarten*" ("The Vineyard") on the recently-developed *Seestraße* in Thalwil. Her well-situated family lives in a thriving community and at an optimistic time: The village, surrounded by meadows, orchards and vineyards, has developed rapidly following the construction of the *Seestraße* and the opening of the rail line. A series of textile enterprises, such as the *Weidmann* dyeworks, where her father Eduard Keller works as a designer and engineer, provide prosperity and job opportunities. Although Zurich had for centuries been accessible only by boat or over the narrow country road high up above the banks of the lake, the splendid city has now suddenly moved much closer. The "Frogs' Moat" has been replaced by the elegant *Bahnhofstraße*, and the medieval "*Kratzquartier*" between the Limmat, the Lake and the Parade Grounds has been razed and the space made available for expansive promenades. Engineer Keller most certainly took the opportunity to treat his wife Anna (for whom

© Springer International Publishing AG, part of Springer Nature 2018
C. Graf-Grossmann, *Marcel Grossmann*, Springer Biographies,
https://doi.org/10.1007/978-3-319-90077-3_9

the older daughter was named) and the children to the occasional Sunday voyage on the lake or excursion by rail. They admire the *Bürkliplatz*, the *Alpenquai* and the *Quaibrücke* (*Bürkli* Square, Alpine Quay and the Quay Bridge in Zurich), and refresh themselves afterwards with a cup of hot chocolate in the *Confiserie Sprüngli*.

Whilst Marcel Grossmann is attending the Higher Middle School in Basel, Anna Keller, four years younger, is at the Girls' Highschool of the Canon Monastery at the Cathedral. In addition to general education subjects, the curriculum includes modern languages. But in contrast to Mileva Marić, Anna Keller never considers studying at the university level; she finishes off her education with a cooking course. Presumably, she has no further academic ambitions—the foreign co-eds have the reputation in Zurich of being exotic and rather suspect.

These young women, mainly from Eastern Europe, usually wore dark-coloured clothing, high boots and raffish hats; they smoked and acted in general like men, in the opinion of the conservative population. Anna Keller, in contrast, represents the ideal of a girl from a good family, from the tip of her head down to the soles of her feet. She has a strictly upright posture and wears simple, long cotton and linen dresses during the week, with high collars and long sleeves, and an apron, as befits a girl from the country. Her features are regular and clear, her skin is delicate and she has a prominent nose. Her most noticeable feature is her hair; she wears her smooth blonde hair waist-long and braids it or binds it up into an elaborate bun. Her nephew Max Keller and his wife Liesel recall decades later in a witty and surely not quite objective manner how the relationship of the young couple began: "*Once upon a time, in Thalwil on the lovely Lake Zurich, there lived three lusty boys. They were handsome, well-groomed and brought up, intelligent and, in particular, very clever! They had in fact discovered two especially charming girls down on the banks of the lake in the fine House of the Vineyard, who were growing up, with their two brothers, to be lovely young women. [...] One of the three Grossmann brothers, Marcel, went through school with joyful zeal and followed his path unswervingly; he became a friend and helper of* Albert Einstein, *with whom he launched the Theory of the Relative in Zurich* [70]."

Four photos Anna's mother, Anna Keller (1920|Hedwig and Henriette Grossmann (1921)|Hedwig Keller (1915)|Anna's father Eduard Keller (1920)

Photo The house in the *Holderstraße* (now *Hölderlinstraße*), Zurich

The charming girl from Thalwil marries her Marcel when she is 21. We have no way of knowing what impressions she took away from her stays in Frauenfeld and Basel, the first stations of her married life. But it is certain that she was happy to live in Zurich. She spent her whole life there, with the exception of the first years of her marriage, and so was able to follow the development of the city from close by. She watches as the *Kunsthaus* ("House of the Arts") is built and inaugurated in 1910. Opposite to it on the *Heimplatz*, the "*Volkstheater am Pfauen*" ("Public Theatre at the Peacock") morphs from a temple of amusements—where in addition to farces and burlesque, snake dancers and fakirs perform—into a serious theatre of the spoken word [71]. And Zurich offers not only nourishment for the spirit; Anna finds here all the ingredients that she needs to prepare excellent meals for her family and their guests. She is a talented hostess, and the Grossmanns enjoy inviting company and often do so. Assistants and students come on *jours fixes* to visit and are received with tea and Viennese confectionary like the "*Kuglerli*", and colleagues and their wives are frequent guests. Once a month, faculty wives from the ETH meet for tea at Anna's. Often, professors and *patres* who are passing through will make a stopover at the Grossmanns'. On grand occasions, a "*Kochfrau*" ("kitchen helper") is hired, and she serves the meal wearing white gloves. Later, such festivities become rare, owing to her husband's illness and the lack of money, which is regretted by all; more and more often, recipes like the "cheap cake" from her cookbook, written in her own hand, are deployed. But Anna will not be deterred from pampering at least her own family in a culinary fashion. Her son Marcel, for example, always has a lavish, lovingly decorated *Linzertorte* (Linz cake) for his birthdays.

Anna Grossmann sews her own clothes, at first by inclination, and later by necessity, and she wears them with a natural dignity. One wonders whether she practised an upright posture as a young girl by walking around with heavy books balanced on her head, as was often recommended in those times? She suffers under her occasional nickname '*Stolzgüggel*' ("proud as a rooster"), since she believes herself in fact to be rather shy. Shy? Well now, her daughters-in-law will later not necessarily consider her to have been shy; she can be—in spite of all her goodness and warmth—rather distanced and strict. Anna, who is not unhappy to hear herself called "*Frau Professor Grossmann*", is disciplined and aware of her status, and she demands of herself and of those around her that they exhibit a generous portion of self control and strength of will. Without those qualities, she would have no doubt been shattered later by her tasks, since her life will take a quite different course from what she has imagined. In the beginning, however, all appearances surrounding her wedding in 1903 point to the start of a generally happy married life and family.

In the picture taken at the time of their engagement, both of the young people are looking gracefully in the same direction. Their temperaments could not have been more different. Marcel is expansive, eloquent and generous; Anna is in contrast quiet, prefers to keep her opinions to herself, and is careful how she spends her money. Whilst he consciously or unconsciously draws on the upper middle-class lifestyle from his early years in Budapest, she represents the protestant, down-to-earth and thrifty spirit of a country girl. We know of no serious disagreements between the husband and wife, perhaps because their areas of responsibility were so different, and—as was usual at the time—Marcel serves as the breadwinner, and Anna as housewife and mother. She proudly witnesses how the sphere of influence of her husband grows, how he blooms at the ETH and within the circle of his professional colleagues and friends. She makes no protests when her husband once again spends long nights brooding over his theories, or when he is short-tempered at the dinner table, and his thoughts are obviously elsewhere. When Marcel Hans is born in 1904, and Elsbeth in 1909, she takes loving care of the two offspring, with the help of a nurse. The children spend their carefree youth in the *Herrenbergstraße*. They both attend elementary school at the neighbouring *Oberstrass* School. Already in this rented apartment, and especially later in their own apartment in the *Holderstraße*, Anna demonstrates her good taste and sense of style.

Their own apartments—their own house: The two brothers make their dream come true. On December 18th, 1919 [72], the time is ripe; one half of the stately multiple-family house at *Holderstraße* 14 (today the *Hölderlinstraße*), with its own area and a small garden, now belongs to them. They will live there in future with their wives and children. Their wives are Anna and her sister Hedwig, eleven years younger. Eugen and Hedwig, who were married in April 1916, live on the first upstairs floor, whilst the older couple live on the second.

Two brothers become acquainted with two sisters in Thalwil, marry them, and they all live under the same roof—will that turn out well? At first, everything points in that direction. The brothers move in one after the other; first Marcel, and then, in the fall of 1920, Eugen. They imagine their common lives in that bourgeois and cultivated area of the city, *Hottingen*, in the most rosy colours: The two men get along splendidly, and Anna and Hedwig have known each other from earliest youth; there will be a lively exchange between the two apartments on the first and second floors, and the house is also well-suited for children. Marcel Jr. and Elsbeth are by now 16 and 11 years old. Eugen and Hedwig have a little daughter named Henriette, who is born in 1920. Later, as an adult, the 'baby' describes the situation in the *Holderstraße* as follows: "*The idea was well meant, but it didn't turn out so well. As much as the two brothers loved each other, the two sisters had nothing at all in common. My mother was and remained a very spoilt and selfish girl, had a very comfortable and easy life, and wanted always to play the prima donna* [73]."

Die glückliche Geburt eines gesunden Mädchens

Elsbeth

zeigen ihnen mit grosser Freude an

Marcel und Anna

Grossmann-Keller

Voltastr. 29

Zürich V, den 22. März 1909.

Photos Elsbeth and Marcel Hans Grossmann (1915)|Birth announcement of Elsbeth Grossmann (1909)

Even though in his memoirs, Eugen portrays his wife in a much more positive light, and places her modesty and sincerity in the foreground, one cannot help but notice that Hedwig was the pampered pet of the family; she is used to having her wishes promptly fulfilled. Hedwig is not a classic beauty like her sister, although she has a striking face and is always elegantly dressed. She admires Anna, and at first, life in their shared house is lively and harmonious. Both couples decorate their apartments in a manner corresponding to the *belle epoque* style of the house itself. In Anna and Marcel's apartment, dark silk wallpaper, wood panelling and parquet flooring contribute to the fashionably sombre atmosphere in the drawing room. The generously proportioned entry hall has modern linoleum flooring, which is waxed to a high gloss each week, the pride of the housewife. The furnishings are heavy and comfortable; some of them date from Father Jules' family home in Mulhouse, for example a secretary, a commode and a table. The sofa and the armchairs, upon which Albert Einstein is especially pleased to sit during his visits, are still among the family possessions today.

Eugen and Hedwig live in a more expansive style: they have a cook who also doubles as a maid, a nurse for their daughter, and a seamstress who comes in once a month; but Anna and Marcel have only a wry grin for the extravagance of their younger siblings. Hedwig loves parties, for which she engages a lady with a white lace headdress to serve at the table. She also enjoys shopping and walks daily to the centre of town at a heady pace, her breathlessly panting daughter in tow. There, they spend an hour in the shops, drink a cup of hot chocolate in a café, nibble on some pastries and then march briskly back to *Hottingen*. The younger couple treat themselves to elegant holidays, travel as their hearts desire, and often spend their summers on cultural and gourmet tours in Italy. Little Henriette is sometimes left behind in Zurich, and Hedwig is on occasion quite astonished to find that her sister is not always willing to look after her daughter. It would to be sure never occur to Hedwig to offer her help to Anna in return.

For ten years, the lives of these two quite dissimilar Grossmann families proceed for the most part harmoniously. But the dream of a happy life together in their shared house gradually fades away unnoticed.

10

Publications

Let us turn back the calendar by a few years, to the time before the Grossmanns moved into the house in the *Holderstraße*. At the outbreak of the First World War, the lives of many citizens, even in Switzerland, were disrupted. Men of an age for military service are called up to guard the country's borders, so that everyday tasks fall to the women, children, and older men. The beginning of August is a busy time for farmers, and all available hands are needed to bring in the hay and grain and to put up supplies for the winter. In the schools, most of the teachers are now women, and in hospitals, the lack of doctors leads to long waiting times, whilst factories are at a standstill.

Marcel Grossmann was not called up. Owing to their overly small chest measurements, he and his brother are deferred from military service, which saddens and disturbs them, as one can read in Eugen's diaries. Grossmann wants to make himself useful in his own way. The fate of the students who have been called up concerns him; nearly 600 Swiss students from the ETH are on active duty, whilst 180 foreign students return to their homelands at the beginning of the war to fight for the Central Powers or for the *Entente* nations—a paradoxical situation for young men who were sitting together in lecture halls shortly before. For 15 of them, the war brought an untimely end; the sad résumé speaks for itself: fallen on the Western Front, in the Balkans, in Galicia, in Flanders, in the Champagne, at Isonzo... also, a number of students are taken prisoner of war.

© Springer International Publishing AG, part of Springer Nature 2018
C. Graf-Grossmann, *Marcel Grossmann*, Springer Biographies,
https://doi.org/10.1007/978-3-319-90077-3_10

The case of the doctoral student Herbert von Wayer can serve as an example for many of their fates: This young man, from Pula (in what is today Croatia) had come to the Polytechnic in 1908, where he obtained his *Diploma* as a Specialised Teacher of Mathematics and Physics in 1912. He has to interrupt his studies several times in order to perform his military service in Austria-Hungary. Wayer is lucky and survives the war; he returns as a doctoral student to the ETH and becomes Grossmann's assistant [74]. Other young men remain for years in prisoner-of-war camps. Elsbeth Grossmann recalls later how her father had organised "*Prisoner-of-war and Students' Aid*", by writing articles and submitting statements to the offices in charge of prisoners, and by collecting teaching materials for prisoners and sending them to the prison camps.

Whether in service or deferred, for a whole generation of young men, the war causes a deep break in their development and careers. But their attention is often directed toward their own personal destinies and those of the people around them; they spend little time thinking about the background and causes of these events. In the *Neue Helvetische Gesellschaft*, there is much talk that the schools and families should do more to educate young people to be responsible citizens. Marcel Grossmann also finds that his students and assistants lack an interest in politics and in serving the community, and he wants to increase their awareness. In the first morning edition of the "*Neue Zürcher Zeitung*" on August 13th, 1915, his lead article deals with a lecture that he gave to the Zurich Group of the NHG. In line with the goals of that organisation, Grossmann has formulated his thoughts on the responsibility of the institutions of higher education, i.e. the preparatory schools and colleges, in the field of "national education". The mathematician suggests that these institutions should more intensely promote the patriotic sensibilities of their students. That this is not possible owing to the "*increasingly threatening expansion of the material to be taught*" is regrettable, and he suggests that the pressure be reduced "*by a serious reduction of the material to a tolerable level*". Only in such a way could "*an intensification of education be achieved, namely in both national and in general-humanistic terms*". He is convinced that this will be possible without endangering the traditional role of the schools. To this end, he wants to offer "*instruction in the history of our nation up to modern times, as well as a lively dissemination of civic skills*", which should not be simply "*a dry required subject, but rather a convincing plea*" [75].

Three years later, he makes his own special kind of plea: In the socially disturbed and economically difficult times at the end of the War, as his brother Eugen Grossmann reports, *"several intellectuals, at their forefront Professor Egger and my brother Marcel, initiated the attempt to found a new daily newspaper"*. The epitaph written by Louis Kollros gives some indications of the motivations for this journalistic and financial adventure: *"He [Marcel Grossmann], together with Professor Egger from the University of Zürich, founded a courageous and independent newspaper [76]."* Courage and independence have perforce become foreign concepts for the media during the war years. The historian Georg Kreis dealt extensively with freedom of the press during this time, and he found that the Swiss national government had limited these constitutionally-guaranteed rights of the press already in 1914. A preliminary censorship and a further decree in 1915 *"justify the suspicion that they wanted to suppress every sort of criticism of the army, especially of its activities in the border regions"* [77].

Photo Marcel Grossmann (1918)

These restrictions are indeed lifted again after the end of the War, but it is conceivable that the journalistic muzzle of the previous years had awakened the wish on the part of Marcel Grossmann and his fellow campaigners to have their own, genuine public voice. Together with August Egger, a legal expert and professor of Civil Law at the University of Zurich, as well as Albert Stroll, Marcel Grossmann forms an initiation committee, and on December 19th, 1918, the *"Verlagsgesellschaft Neue Schweizer Zeitung"* ["Publishing Cooperative for the New Swiss Journal"] is entered into the Commercial Register. The Administrative Board is presided over by Marcel Grossmann, whilst Dr. Wilhelm Nauer is Director. Albert Stoll is Delegate and Paul Alther-Kürsteiner and Hugo Bartholdi, a nephew of Anna Grossmann's from Thalwil, are Members. The seat of the Cooperative is initially at *Neue Beckenhofstraße* 47 in Zurich, District 6; two years later, it has moved to *Pfingstweidstraße* 57. To be a Member of the Cooperative, one must be Swiss and acquire a share in the Cooperative valued at 100 Francs or more.

This new paper leaves no doubt concerning its patriotic standpoint; its name is its programme: The *"Neue Schweizer Zeitung"* strives to analyse the politics, economy and culture of the post-war Swiss society, independently of political parties and from a new, original orientation; and to support political interest, stimulate economic life, and to work against *"foreign economic infiltration"*. In the cultural area, it aims to *"support good 'Swissness', to strengthen and protect [Swiss culture]. Both in terms of politics and economics, as well as in terms of culture, the new paper is intended to be an organ of free discussion* [78].*"*

An ambitious title and an aspiring undertaking, especially since in Switzerland, already a large number of established newspapers of every political complexion are being published. Eugen Grossmann is dubious about this enterprise and his advice is to publish a more modest weekly or monthly journal instead of the planned daily newspaper; but the enthusiasm of the initiators cannot be dampened. On December 20th, 1918, the moment has arrived: The first issue of the *"Neue Schweizer Zeitung"* appears. The editorial staff, led by the Germanist and publicist Dr. Hermann Schoop and by Dr. Werner Ammann (known to Marcel Grossmann as a member of the NHG), publishes at first two, then three issues per week. Brother Eugen, in spite of his initial doubts regarding the project, declares his willingness to become a member of the Editorial Board. It is responsible for the selection of editors, and if desired, its members can also write articles for the paper. In the coming years, a number of articles will be written by the two brothers.

The start of the paper is encouraging: right away, 6000 subscribers are recruited. The authors are granted great editorial freedom and can take up a wide variety of topics as they wish. Politics, Economics, Finance, Science, Society, Culture... the palette is colourful and variegated. Reports from other newspapers are also taken up and commented, and the personnel of the newspaper engross themselves in their columns with gusto. Marcel Grossmann applies himself in particular to scientific and political themes. He writes of the fascination of mathematics, physics and astronomy, of the achievements of science and of the new, optimistic world view which has opened up because of them. In June of 1919, he points out proudly that the Theory of General Relativity has been given support by the results of observations of the total solar eclipse on May 29th, 1919.

His articles [79] have a personal note and are quite trenchant, as some examples demonstrate: On January 28th, 1919, under the title "*Zur Aushungerung Deutschlands*" ["On the Systematic Starvation of Germany"], he writes a committed plea against the blockade of the neighbouring country, which would lead to the death by starvation of 700,000 human beings. A year later, during a lecture by Albert Einstein in Berlin, there is a great commotion because the physicist announces that he plans to give lectures for the general public. Whilst German professors and students harass and berate Einstein in a petty way because of his political stance—and likewise because of his increasing renown—Grossmann directs his "*words to German colleagues*" and warns of the damage to the reputation and the future of Germany which could be caused by such a "*reactionary spirit*". He strongly supports a science which can develop independently of any political apportioning of blame. Six months later, he again expresses his shock at the "*anti-Einstein agitation in Berlin*" and the growing antisemitism, which has gained entry into academic circles, as well. "*That, however, the world of German scholarship has not energetically opposed such a shameful agitation, is exceedingly sad and very serious.*"

Alongside science, it is politics which occupies the attention of the author; more than once, he joins the debate over the fate of the country of his birth, Hungary, following the First World War. After the collapse of the Danube Monarchy in October of 1918, Hungary was divided up among the Slovakians, the Romanians, the Croatians and the Hungarians, signalling the end of the multi-ethnic nation. In April 1920, under the title "*Ungarns Schicksal*" ["The Fate of Hungary"], Grossmann writes "*Our Swiss daily papers are filled at present with the laments, protests and calls for help from Hungary; they are worthy of a closer look at their justification. The new government is trying to wipe away all traces of the past war and the revolution. [...].*

It is attempting to give Hungary back 'its thousand-year-old boundaries'." He judges these 'major-power ambitions' of the postwar government in an extremely critical fashion: "It is [...] permissible for a Swiss citizen who was born in Hungary and is well acquainted with the situation there from his own observations, who was always very sympathetic towards the Hungarian people (if not their leaders and politicians), and who showed those sympathies to Hungarian prisoners of war, to examine this question from his own standpoint." And he does so with clear and sharp words: "Millions of foreign-born and still foreign subjects—Serbs, Romanians, Slovakians, Germans, etc.—have suffered under low-grade rights for a very long time, and right up to the catastrophic end of the double monarchy. They were kept down in every possible way by the 'Master Race', the Magyars: culturally, politically, and economically. One has to have seen for oneself how the Slovakian villages in the Tatra looked shortly before the War, in what primitive and uncultured circumstances that ethnic group were forced to exist, how they were looked down upon. During the War, a reign of terror raged in Hungary, 'after the German pattern, but interpreted congenially and varied'." Marcel Grossmann strongly supports the allied peace treaty with Hungary, and he ends with the words, "The new government is working purposefully to re-establish the reign of the nobility, the military, and the monarchy. The Hungarian people however, as far as they are allowed to express themselves, would not appear to be ripe for democracy, in spite of the fearsome lessons of the War, which it owes to its previous leaders. The only positive aspect is that this reactionary development will have to come to a halt in the face of those peoples who have a right to freedom after a millenium of repression."

Another year later, he expresses his thoughts about the governmental system 'democracy', and gives a very current report of the conference of the Neue Helvetische Gesellschaft, only four days earlier. "Is now democracy in fact the best form of civilisation? Theoretically, for certain. But one can also quote reasons why a monarchy or communism could also be considered theoretically to be the best systems. That is not the deciding factor; it is rather the results of practical experience. And there, an attentive and critical observer will not have missed the signs of a hypertrophy of the real basis of democracy: Exaggerated equality in the interest of political freedom undermines personal freedom, leading to 'etatism', that anonymous form of tyranny."

Although the articles in the "Neue Schweizer Zeitung" generally exhibit a high level of journalism and are carefully written, in spite of their occasional subjectivity, the financial problems of the newspaper become more and more serious after three and a half years. The paper suffers under competition from other media, and under the difficult economic situation in Switzerland; its advertising section becomes thinner and thinner, and the number of subscribers is shrinking. The interest of its readers in topics that were discussed with engagement and controversy in the beginning is now waning.

On May 13th, 1922, an extraordinary General Meeting of the Cooperative is called; its central topic is the future of the newspaper. A month later, the editorial management then announces that the *"Neue Schweizer Zeitung"* will cease publication at the end of the month [June], since *"the external difficulties opposing its continued existence have proven to be insurmountable"*. These are in particular financial problems. As August Egger mentions in self-critical fashion in his epilogue, the paper had started out with an overly slim financial base and had too little financial and professional expertise. The enterprise had indeed begun in relatively good form, but it could not survive the *"serious economic crisis which struck our nation"*. A certain amount of defiance can be detected in his tone when he writes that the newspaper *"was right from the beginning in contradiction to many opinions which predominate in this country, and it was swimming upstream. Many were not able to tolerate such an independent paper. Many also failed to understand it, and our attempts to make our goals understandable were met in some quarters with insurmountable difficulties [80]."* The Cooperative, however, is not immediately liquidated, so that this group of Swiss citizens could remain intact. The communication closes on an optimistic note, asserting that *"the time will come when the need, in whatever form, for a politically independent medium which does not represent any particular class interests, and which in terms of foreign policy consciously promotes the democratic ideal that is directed toward supporting the League of Nations and international cooperation, will make itself felt as an imperative"*. This in fact never happens; the Cooperative is disbanded in 1926 and is deleted from the Commercial Register. Of the initial capital, which amounted to 120,000 Francs soon after the founding of the Cooperative, only 500 remain following its liquidation; the shares have become worthless.

Marcel Grossmann has invested his lifeblood, his time and his capital in the *"Neue Schweizer Zeitung"*, and this fact permits a deep insight into his views and convictions. He considers himself to be a patriot, even though he and Eugen never regarded themselves as typical Swiss, owing to their youth as Swiss Abroad and also to their Alsatian father. The mathematician supports the extreme Helvetian views of the newspaper, but he warns against nationalism and the dangers which accompany it. He speaks in favour of international cooperation and relations and is an enthusiastic supporter of the League of Nations which was founded after the First World War, and of membership in it on the part of Switzerland. He makes no secret of his belonging to the *Neue Helvetische Gesellschaft*, reports on its meetings and on the fact that the opinions of its members are often strongly divergent. All in all, Grossmann appears to have been a wide-awake, critical contemporary of that time, who campaigns vehemently and passionately for geopolitical concerns. Indeed, sometimes his prejudices also show through, for example in his statements about the effects of the War on Austria-Hungary, although he has no longer visited the country of his birth for a number of years; but he analyses many problems correctly and has thoroughly realistic views of the political developments during the years between the World Wars.

This period is characterised by economic and social problems: The end of the First World War leaves Europe bled dry. Even Switzerland, which was spared from active participation in the War, suffers from shortages of raw materials. After the European currencies had been stable for decades, the War began a period of destabilisation for the Swiss Franc [and for other European currencies]. Prices began to rise in 1916, buying power decreased, and the economy suffered under the hyperinflation of its neighbours. In addition, there is social unrest: Workers' rights are defended in a voluble manner due to the increasing strength of the socialist party and of the communists, and strikes become a common occurrence. Eugen Grossmann recalls later in his memoirs that the social turmoil was great, "*although the workers had almost generally obtained a 48-h work week with complete wage equivalence*". The coming years will be an acid test for democracies, and in Switzerland, the dialogue between employers and employees has to be learned from scratch. Marcel can now no longer publish in the "*Neue Schweizer Zeitung*", but he remains preoccupied by the questions which it had raised.

11

Affinities

During the 1920's, Marcel had arrived at the zenith of his career—he is extremely active both professionally and privately, and is engaged on many fronts. This is no doubt the reason why—unlike his father, his brother, and his son—he never kept a diary; he simply did not have the time to do so. And later, when the hours pass by with painful slowness and he would have had time, he no longer has the inclination.

After Grossmann and Einstein had lost contact with each other during the First World War, their exchanges begin again when it is over. Einstein is still living and working in Berlin. After a few years of bitter disputes with Mileva, she had finally agreed to a divorce in early 1918. Following the end of the War, academic exchanges across international boundaries are once again established. The physicist Einstein goes to Zurich to give a four-week series of lectures. The two friends see each other again a year later, when Einstein travels to Switzerland in the summer. He is now free to marry his cousin Elsa, but he is struggling with various kinds of problems: his health is causing him trouble; he is suffering from an oversensitive digestion. His former wife Mileva holds the separation against him, and his sons Hans Albert and Eduard complain that he pays them too little attention. And the rapidly increasing inflation in Berlin makes it more and more difficult for him to keep up alimony payments to support his family in Zurich.

The concrete motive that caused the two friends to again take up an intensive correspondence was a remarkable situation surrounding the Western Swiss physicist Edouard Guillaume, cousin of the later Nobel Prize winner Charles Edouard Guillaume. Einstein has known him since his

© Springer International Publishing AG, part of Springer Nature 2018
C. Graf-Grossmann, *Marcel Grossmann*, Springer Biographies,
https://doi.org/10.1007/978-3-319-90077-3_11

days in the Patent Office; Guillaume has translated his articles into French on several occasions. Around ten years later, Guillaume starts a campaign against Einstein and his theory, based on his own lack of scientific understanding. Guillaume's statements are indeed challenged in a critical and differentiated manner by other colleagues, and their loyal and for the most part objective opposition is appreciated by Einstein; but as for Guillaume, it is increasingly evident that he has simply not understood the theory. He in fact corresponds with Einstein and informs him in advance of his objections, and Einstein shows a notable patience towards him and attempts, unsuccessfully, to convince him; but Guillaume clings to his opinions: He rejects the Theory of General Relativity, because it does not deal with the concept of a 'universal time' [81]. Grossmann notes with increasing discomfort that Guillaume's abstruse ideas are finding a growing acceptance among certain Swiss scientists. Even Michele Besso, Einstein's loyal companion, seems to be warming to Guillaume's arguments. On September 9th, 1920, Marcel writes to Albert:

Dear Albert, In the same post I am sending you a paper by Mr. Charles Willigens from the '*Archives des sciences physiques et naturelles*' [Archives of Physical and Natural Sciences]. As you can see, a cult is forming around Guillaume that thinks it must correct essential points of your concepts. Even though this matter is unlikely to be of interest to you personally, I think that it would be in the interest of relativity theory if you had a brief joust with Guillaume someday, perhaps as a short article for the 'Archives', for which I would gladly provide the translation; or else you could simply state your views in a letter to me, the scientific gist of which I could pass on to the 'Archives'; that would please our sympathetic colleague Guye very much.

There is a danger that from the unchallenged appearances by Guillaume and his disciples, which have also appeared in the daily press, dissemination of the fundamental ideas of General Relativity will suffer harm in the French-speaking region, where the public is all too quick to embrace the superiority of everything French, even in this field. All the more since the depraved campaign against you in Germany has also found an echo here. I therefore believe that I am justified in asking you to inform me in outline for what reasons you reject Guillaume's ideas!

I very much hope that you and your family are well and happy. Both our boys, who are in the same class in high school, are already calculating with logarithms. We are also doing well, after my dear wife withstood a nasty sepsis just a year ago, which brought her to the brink of death. But now she is up and about again and more cheerful than ever. Are you still not ripe for Zurich yet?

With hearty greetings to you and your wife, your sincere friend, M. Grossmann [82].

The sepsis mentioned in the letter was a serious illness which both Anna and her daughter suffered in 1919, quite possibly a result of their having been infected with the Spanish Flu. Elsbeth also suffered from tuberculosis, from which she recovered after a high-altitude cure in the *Bündnerland*, in Arosa. There, she also received radiation treatments using a quartz lamp. That treatment was however still in its early stages, and her neck was burned by the ultraviolet radiation; she carried the scars for the rest of her life. Back to the Guillaume affair: Evidently, Marcel's appeal had the desired effect; Albert answers by return mail and sends Marcel his statement for publication. It is however clear from his letter of September 20th, 1920, that he finds the whole situation rather amusing:

Dear Grossmann! This world is a strange madhouse. Currently, every coachman and every waiter is debating whether relativity theory is correct. Belief in the matter depends on the political party to which the person belongs. Most amusing though is this 'Guillaumiade'. For in it, someone has for years on end been propagating the saddest nonsense in scientific jargon to the illustrious experts in the field, and this with impunity, without being reprimanded. This shows us quite clearly how the judgements and values prevailing amongst the flock of scholarly sheep rest on the narrow foundation of a few discerning minds.

Refutation is not such an easy matter, if one is not even in a position to understand the assertions of their supporters. I took every trouble; I have thought it over at length, have carried out a long correspondence with Guillaume, but I met with nothing besides mathematical symbols devoid of any sense; a factual reply is out of the question; one can only express an opinion. I enclose one for the '*Archives*' with this letter.

You ask me in your witty fashion, 'Are you still not ripe for Zurich yet?' To that I can say the following: In personal terms, it is wonderful for me here. My immediate colleagues are truly welcoming and friendly. The Ministry does everything it can to read my wishes from my eyes and fulfil them. Truly dedicated friends are also not lacking, either. But it is becoming increasingly difficult for me to support my family in Zurich; it would have long since been impossible, if not for the aid of special circumstances, which are however not guaranteed over the longer term. Transferring my children to Germany is in my opinion not a proper alternative. So it could happen that I will have to consider giving up my present position owing to these external reasons. But I dread taking that step, since they would resort to the most desperate measures to keep me here; not so much because they value me for personal reasons or for my brainwork, but more because I have become a kind of idol due to all the publicity. I play a similar role as the relics of a saint which absolutely have to be kept in the abbey church. My departure would be perceived as a lost battle. It would be damned difficult for me to mobilise the necessary rigour, even though it may become necessary. And I also believe that they would always raise the required cash in case of necessity. The tragic aspect of my situation is that I can't muster even a small fraction of the self-esteem that I would need in order to play with 'dignity' this role that I have been assigned through no fault of my own.

I am genuinely happy to hear that your wife is again healthy and content, and that you are all enjoying life. I am no less happy that our lads are in the same class in school, just as we were. Let's hope that we will soon see each other again This year, I am inviting my boys to come to Germany (in October), because a trip to Switzerland would be too costly for me.

With hearty greetings, Yours sincerely, Einstein [83].

The attachment to the letter that he mentions is in a class of its own; Einstein indeed did not mince his words:

For the 'Archives':

In recent years, Mr. E. Guillaume has repeatedly published statements about the Theory of [General] Relativity in this journal, and he has in particular attempted to introduce a new concept ('*temps universel*') into that theory. At the repeated prompting of the author himself as well as from other colleagues in the field, I believe it to be necessary to declare the following.

In spite of having spent considerable effort on this matter, I have not been able to find any kind of clear sense in Guillaume's expositions. A correspondence with him carried out with utmost patience has also not allowed me to arrive at that goal. In particular, it has remained completely unclear to me just what that author means by the term '*temps universel*'. My ability to comprehend has not even arrived at a point which would permit me to formulate a substantive rebuttal. I can instead only express my conviction that there is no kind of clear chain of reasoning behind Guillaume's statements.

For Marcel, Albert adds a note: "*Dear Grossmann! Please request that the 'Archives' send the proofs to Guillaume for correction. My comments are harsh, but I have found no other way; this nonsense has just gone too far* [84]."

In the end, this text is not published; Einstein's friends evidently find it too hurtful. Grossmann publishes a somewhat more diplomatic statement under his own name. But the academic dispute between Einstein and Guillaume continues to smoulder for quite some time and takes on increasingly cynical tendencies on the part of Guillaume. Grossmann is also pulled into the firing line of the critics, owing to his uncompromising loyalty to his friend. After Einstein has written to Guillaume again on December 16th, 1920, and has pointed out to him with angelical patience that he "*has not achieved an understanding*" of Guillaume's articles "*in spite of zealous efforts*", his correspondent misunderstands the fine irony. He assumes that Marcel Grossmann has been exerting a negative influence over the physicist. In his reply on December 23rd, 1920 to Einstein, Guillaume makes fun of Grossmann, who had taken a clear position on the Theory of General Relativity at a mathematics conference in Straßburg the previous September:

> My dear Einstein, No! I am not annoyed, but rather saddened, for it is always sad to see an excellent person such as yourself who is in the process of committing an error in such a crude fashion. You have taken on a fearful responsibility, and I hope that its consequences will not be all too dreadful for you! Your friend Grossmann is playing a strange role in all this, and the little speech that he made in Strasbourg after my lecture attracted considerable notice. We are not accustomed in the jovial atmosphere of the mathematical world to hear such words; there, where only objective argumentation counts. One could laugh oneself to death on hearing in a mathematics conference that 'an identity is uninteresting for mathematicians'! [...] [85].

Later, Einstein loses track of Guillaume and his confused ideas, not least because he is now spending much of his time on professional travels. In 1921, he visits the USA for the first time and also goes to England. In October of 1922, he leaves on a trip of several months, which this time takes him to Japan and, on the return trip, to Palastina and Spain. On the outbound journey, Einstein and his second wife Elsa board a ship to Japan in Marseille, and send Marcel a postcard from the *Exposition Coloniale* there. On its front side, the *Chateau d'If* is pictured, the famous island prison from which the Count of Monte Christo is supposed to have escaped.

Dear Grossmann! Seek him not in the college course, / Seek him by a glass of Tokay-er. / Seek him not in Hedwig's Church, / Seek him by Mademoiselle Maier—This is Heine. But I sought you in vain at the Poly[technic] and did not have any more time to look for you at your roost. So, here's another hearty greeting in case I drown or am wrung out to the point of disintegration from all that yapping. Yours, Albert.

Congratulations on junior [filius] [86].

The congratulations are for Marcel Jr., who is in the meantime 18 and who graduated with his school leaving certificate without problems in September. In French, History, Natural Sciences and Chemistry, he received the best marks (6); in German, Latin and English a 5, and in Gymnastics a 3.5 ... the family traditions were thus maintained.

Shortly after Einstein sends off the card from Marseille, the academic crowning of his career occurs; he receives the Nobel Prize in Physics for 1921, after having been nominated a number of times in earlier years. Although he knew that he would have an excellent chance of receiving the coveted prize this time around, he travels unaffectedly to Japan with Elsa and learns of the committee's decision on board the ship bound for Tokyo. Will he interrupt his trip to Japan in order to be in Stockholm on December 10th, the traditional day of the prize awards? Hardly! He calmly continues his stay in Japan and doesn't even mention the sensational news in his diary. Considering how much honour is connected with the Nobel Prize, the coolness and nonchalance with which he writes to his friend in November 1922 from Tokyo are astonishing. The postcard shows a section of the cozy Hibiya Park in Tokyo.

Dear Marcel! I am sitting here next to Fucisava in Tokyo (but not really in Japanese style, rather on chairs) and thinking with friendship of you. What have I experienced in these two months! You would have stared, sweated and grinned at many things if you had been along. We made a grand entrance at every stop. Hearty greetings to both of you from Einstein [87].

Photocopies Postcard (both sides) from Albert Einstein to Anna and Marcel
  Grossmann (November 23rd, 1922)

The prize money of 121,572 Swedish Crowns is a highly welcome financial boost. This stately sum allows Einstein to fulfil his promise of material
support made four years earlier in his divorce agreement. Mileva uses the
money to acquire properties. This means one less worry for Einstein. His

fame grows inexorably, and he is bombarded from all sides with requests and offers. Even in these labour-intensive years, the two friends maintain their correspondence; but now their letters are more concerned with world politics. From 1922 to 1923, Albert is a member of the League of Nations Commission. In a much-noticed public speech, he declares his resignation, but later returns to the Commission. Marcel tries to convince him of the usefulness of the young organisation. From Albisbrunn, where he is taking a cure, he pleads for patience: "*It's demanding too much to expect the young League of Nations to clean up the chaos that the injudiciousness of the last few years has created* [88]." His letter is reprinted on pp. 132 ff. On December 28th, 1923, Albert answers him amusedly from Berlin:

Dear Grossmann! I am always delighted with any sign of life from you, whether you are giving me a lecture about the League of Nations now or telling me anything else. Nothing will come of the lecture. Because, first, I don't want to go on such a major trip yet again, and second, I have nothing to present to the wild public at large. What you write about my filius pleased me very much. One sees from one's own children how one is gradually maturing into a veteran. I don't regret my behaviour toward the League of Nations one bit. It is simply a tool of the party in charge without any independent power; and it doesn't look at all as if anything really valuable could come out of it. If it must languish on, it can just as well carry out little individual functions to reduce tensions. It just discredits its own true goal due to its lack of energy and goodwill. I'm glad not to have anything to do with it. Scientifically, I have just found a very interesting possibility for perhaps(!) doing justice to the facts of the quanta, as seen from relativity theory; I want to send you the article when it's printed. If only the pursuit of this thought didn't run up against such infamous mathematical obstacles! Now it seems as if the [gravitational] redshift of spectral lines is also finally becoming true, despite the existence up until today of very weighty doubts about it put forward from the observational side. St. John at Mount Wilson, who has hitherto been the most sceptical, now regards the effect as verified on the grounds of very comprehensive and careful measurements. Thank you very much for wanting to preserve the option for me of coming to Zurich. But I am being treated so well and solicitously here and I have it so good that it really would be mean if I left. Besides, theoretical physics is so exceptionally well represented by Debye and Weyl in Zurich anyway, as perhaps nowhere else in the world.

Wishing you and yours a happy 1924, your A. Einstein [89].

In 1924, Marcel Grossmann has to request sick leave from his teaching duties for the first time. He continues to work at home and invites Einstein to give a lecture in Lucerne. On January 11th, 1924, he writes to his friend in Berlin:

Dear Albert

Once again I must approach you with an assignment that contains a request. The organising committee for the Annual Meeting of the Swiss Society of Natural Science Researchers would very much like to steer you toward their Meeting. It will take place from October 1st–4th in Lucerne, and they would like to have you give a lecture during the final plenary session, on October 4th. This would thus be a lecture directed at all the natural scientists, and it should deal with some area of physics. I was also asked to request that you name your conditions, in case you are willing to give the lecture. I very much hope that you can and will come; we could see each other, since I plan to attend the Annual Meeting with my wife.

Many thanks for your recent letter; your scientific plans were of great interest to me. I was very happy to learn that the redshift is now accessible [to observations].

Photocopy The letter written by Marcel Grossmann to Albert Einstein (August 1st, 1923). Three pages

Es ist zu viel verlangt, dem jungen
Völkerbund zuzumuten, das Chaos
zu bereinigen, dass der Unverstand
der letzten Jahre geschaffen. Er wird
aber in Zukunft solche Ereignisse
verhindern sollen. Diese Entwicklung
aber zählt mit Jahrhunderten
nicht mit Jahren.

Die Leute, die den heutigen
Völkerbund ablehnen, weil er ihr
(oft frisch lackierten) Idealismus
nicht entspricht, erinnern mich
immer an den Jungen, der da

11-464

Your pessimism regarding the League of Nations is understandable in terms of your point of view; but I still believe that humanity can be helped, if at all, only by following this path. Previously, I thought that the problem of pacifism was so difficult because every generation simply has to again shed its own blood, the lessons of war are so soon forgotten; now, one could nearly be convinced that it will never be more peaceful down here until every individual has gotten shot through the head or the belly!

With kind greetings, from my family also, and to your wife, Yours truly, M. Grossmann [90].

Two months later, on March 15th, 1924, a whimsical postcard arrives from Berlin, and it demonstrates very well the relaxed, witty dialogue between the two friends:

Dear Grossmann! Kascht nüd macha. [Nothing you can do about it.] So I will accept your kind invitation. I guess I shall probably speak about the more recent developments in the fundamentals of mechanics. Now there is only one problem; despite all my searching, I now can't find the letter from Prof. Lugeons with his address (orderliness, blessed daughter …). May I therefore ask you to accept [the invitation to speak] in my name and to make excuses for me, for the reason that I am a slovenly guy, who wouldn't even have passed his examinations without the help of Grossmann's notebooks.

Warm regards from yours, A. Einstein [91].

12

Illness

Indeed, Marcel Grossmann's career runs as smoothly as if on rails. He is a respected faculty member at one of the most renowned technical colleges in Europe, and he has excellent academic, social, political and publicistic networks and relationships in Switzerland and abroad. All the signs point to a long and successful career; but now he can no longer ignore the fact that his health is letting him down.

In 1927, he describes his situation to Heinrich Zangger, a doctor of forensic medicine in Zurich and a friend of Albert Einstein's, as follows: He had always been healthy, apart from his typhoid episode in Budapest; but during a hiking trip of the Natural Sciences Society in the recently-opened national park of the Lower Engadine in 1915, he had suffered an attack of dizziness, and his right hand had lost all feeling and control. Two years afterwards, he had begun to drag his right leg occasionally, and later that had become a steady limp. He had problems with walking and standing, the right half of his body tended to have cramps, and he had difficulties in pronouncing words [92].

For a long time, he fails to take these symptoms seriously, and assumes that they were caused by a minor infection. In December 1920, he asks his colleague Louis Kollros to substitute for him for the first time, since he is suffering from the neuralgic aftermath of a flu, a sore throat or a hefty catarrh. This is embarrassing for him, and he requests that Kollros treat the matter as confidential. As his daughter Elsbeth later recalls [93], he in fact often had a cold during those years, and that was attributed to his frequent visits to the *Grand Café Odeon*. There, "*not only are there brainy discussions, but also everyone is smoking*".

© Springer International Publishing AG, part of Springer Nature 2018
C. Graf-Grossmann, *Marcel Grossmann*, Springer Biographies,
https://doi.org/10.1007/978-3-319-90077-3_12

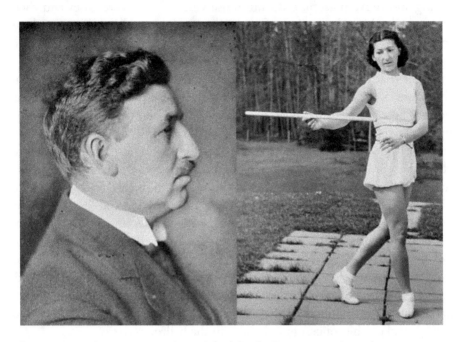

Photos Marcel Grossmann (1930) | Elsbeth Grossmann (1928)

Grossmann knows that he is working too hard and getting too little exercise, that he is overtired and overworked. But who isn't, in the demanding academic surroundings where he works? He consults doctors, many doctors, and each one has his own opinion and comes to a different conclusion. Within the family, also, various theories are passed around. Eugen soon hits the nail right on the head and recognises multiple sclerosis, whose symptoms he knows from the illness of a friend. He recommends to his brother that he move to the country and take walks there in the fresh, pure air. A reasonable suggestion; the air in Zurich can be cut with a knife in winter and on foggy days because of the ubiquitous coal heating.

The invalid however has his own theory and follows it doggedly; he presumes that the vapours which he had inhaled whilst giving his classes in the lecture halls of the ETH during the war years have damaged his health. Because of the reconstruction of the main building, Grossmann had to give his classes from 1915 on in a provisional, much too small and poorly-ventilated auditorium. In addition, the students were perched together and brought their wet winter coats into the room, since there was no cloakroom. The air was so stuffy that sometimes students became nauseated during the lectures. Marcel Grossmann is adamantly convinced that he had suffered a lung and nerve poisoning during those years. Since nerve gases and their effects had achieved notoriety after the devastating attacks with mustard gas in the First World War, it is quite plausible that this knowledge had misled him to his odd theory. Nonetheless, for years he does not doubt for a moment that he will fully recover. Otherwise, he would never have bought the house in the *Holderstraße* in 1919, which is indeed roomy and attractive, but has no elevator. And he wouldn't have moved into the apartment on the upper floor!

In the following decades, Marcel Grossmann leaves no method untried to cure his disorders. He consults one doctor after another, then he tries alternative medicine. Perhaps a stay at a health resort will help? Nearly every year, he goes to health spas and sanatoriums—first to Lamalou and Menton in the South of France, then to Bad Kreuznach near Mainz, and to Degersheim, Herisau, and Albisbrunn in Switzerland. Almost every spring and summer, the invalid spends several weeks at a health spa, sometimes accompanied by Anna. The whole family travels to Albisbrunn on the left bank of Lake Zurich; the lovely park and the spa house make a lasting impression on his daughter Elsbeth, along with the story that Wagner or Liszt was supposed to have played on the concert grand piano in the ballroom there.

In his search for medical help, Grossmann learns of Emile Coué, a pharmacist and psychologist from Nancy, in France. Coué grew up in modest surroundings and was an alert and curious contemporary. In the course of treating his patients, he has observed that the medicines that he sells in his pharmacy do not always have the same effects on the patients; the way in which he describes them has an influence on their effectiveness! When he says, "*With this medicine, you will very quickly recover your health*", the remedy is more effective [94]. Coué begins to study psychology and applies his knowledge in a practical manner. He believes that people can activate and strengthen their own healing powers by means of a positive attitude. Coué thus becomes a pioneer of autosuggestion, and his mantra is, "*Tous les jours, à tous points de vue, je vais de mieux en mieux!*" ["*Every day, in every way, I am getting better and better!*"]. Marcel travels to Nancy in the Lorraine, and his daughter Elsbeth is impressed on his return by the difference in his condition. "*When he left, Papa was hanging heavily on Mama's arm and swung his leg painfully around all along the railway platform. When he came back home, he could walk much more easily for a while.*" For decades, Anna and the two children are firm believers in Coué and resist every sniffle and every illness initially (and often successfully) with a resolute autosuggestion. In the case of their father and husband, however, it can only [temporarily] relieve the symptoms of the illness, whose origin is still not clear.

From 1920 to 1926, the professor is able to continue to give his lecture courses, but he has to ask Louis Kollros to substitute for him more and more often; Kollros proves over those years to be a true and loyal friend and colleague. Grossmann's work becomes more and more difficult for him; walking through the long corridors of the ETH is tiring and burdensome, especially since he often has to carry heavy books or portfolios of drawings. Attending conferences and the general meetings of the societies of which he is a member also tires him unduly. His speech becomes less and less clear, and this depresses him in particular, since he loves polished language and rhetorical brilliance. How humiliating it must have been for him when he had to hold onto the lectern whilst giving his courses, since his legs will no longer obey him; when he stood in front of the blackboard and the complicated formulas that he is writing become more and more illegible; when the students could no longer understand his sentences correctly and avert their faces from his mumbling!

Einstein hears about his friend's problems and is concerned. He writes to Heinrich Zangger on August 18th, 1925 about his trip to Zurich, during which he had met Marcel: "*I visited Grossmann, who has a mysterious nervous disorder with paralysis, and found him better than I had feared* [95]." But this first impression is deceptive. Indeed, the patient's condition seems to stabilise at times—these phases are typical of multiple sclerosis—but then another surge follows. In the winter semester of 1925/26, he again has to request sick leave. The following March, it is agreed that his teaching load will be reduced during the summer semester. That relief is not sufficient; after only the one month of April, during which he can fulfil his duties only partially, Marcel Grossmann is again unable to perform his work.

In 1927, he writes the letter to Heinrich Zangger quoted above (on p. 137). He wants the expert on forensic medicine to confirm his theory of poisoning, so that he will perhaps have the right to an additional compensation from the ETH. Zangger, however, arrives at the same conclusion as have many doctors before him, and he writes to Einstein in somewhat halting German, "*Mr. Marcel Grossmann has asked me to examine him and to write an expertise as to whether he might have acquired his nerve disorder as a result of the poor air and the bad hygienic conditions under which he had to give his lectures for several years, two times 2 h per week, in order to establish a claim to responsibility [of the ETH]. As matters stand, I believe, as you do, that such a proof is impossible; whether an exacerbating effect was present is hard to determine. From what I have seen, he is suffering from multiple sclerosis, which is more on the one side [of his body].*" [96] Einstein is very concerned and answers Zangger on March 31st, 1927: "*I am very sorry about Grossmann. Because of the characteristic remissions in his speech,*

I feared a paralysis. But thank God it is not a question of that, since the process has gone on much too long. I of course do not believe that it is due to the effects of bad air. That is downright naïve." [97]

Zangger is diplomatic; he refrains from giving his patient a final opinion, and instead speaks with Marcel's brother Eugen, who is professor of Statistics and Finance at the University of Zürich. In a letter to Einstein, Heinrich Zangger writes on July 4th, 1927, *"I recently spoke with Marcel Grossmann's brother, since, as I have told you, I was rather embarrassed by Grossmann's request to me to prove that his illness is the result of poisoning. Prof. Grossmann [Eugen] then explained to me that he had not known of that request, but he knew of the general attitude of his brother, and that he himself is convinced that it is a case of multiple sclerosis, and that his brother has already taken early retirement. His brother is now reading and writing a great deal at home, and has accepted the situation. He advised me directly to let the matter drop; thus, I have not yet visited Marcel Grossmann. I will take the liberty of informing him that I have also discussed the whole matter with you in Berlin, and that you will probably be satisfied when you hear of this solution; after all, he can work on what interests him and his disorders will not have such a depressing effect on him."* [98]

Evidently, in the spring of 1927, Marcel Grossmann has accepted that his condition, whatever its origin might be, has made it impossible for him to continue to fulfil his duties as a professor. It must have been very difficult for a strong-willed person with a generous portion of stubbornness such as he was to apply for early retirement. His application was granted without any difficulties on October 1st, 1927. Louis Kollros writes a farewell letter in the name of all the faculty members in his department; it is very hearty, sensitive and appreciative, and was intended to give courage to their sick colleague.

He will need courage, for medical science can give him no hope. The treatment of multiple sclerosis was in its very early stages at that time. Since it is an inflammatory nerve disorder, rest, a quiet life and fresh air are recommended. Whilst today, MS patients are encouraged to remain active as long as possible, and immune modulating medications are available to mitigate and delay the course of the illness (although no cure is yet available), the treatment during the years between the World Wars was limited to relief of the symptoms. MS is often called a painless disease, but this is not true for all of its victims. In the case of the meanwhile retired mathematics professor, it makes itself felt especially through limited mobility, with muscular deficiencies and paralytic symptoms. This paralysis is at least partially due to involuntary muscle cramping [99]. Grossmann also suffers from coordination disturbances; his gait is stumbling and uncertain, because the illness affects the centres in the brain that control motions. He often complains of numbness in his limbs, and feels burning and tingling.

Even though his regular work as a faculty member, his lectures and his travels are no longer possible, at home in the *Holderstraße*, his work continues nearly unabated for years, for Grossmann retains all of his mental powers and is full of ideas. He reads for example of a problem that occurs in the weaving industry. In mechanical looms, so-called cam wheels are used, and they are subject to rapid wear, since they normally cannot be manufactured with the precise shape necessary to reduce friction and wear. At this point, Marcel can profit from his double knowledge of the mathematics of the descriptive geometry of curved surfaces, and of machine design in the textile industry. The son of an entrepreneur and past chairman of the department of mechanical engineering recognises that the mass production of the parts is too imprecise, resulting in their continual reshaping during use. The tinkerer analyses the motion of the cam wheels as a geometrical problem and invents a process by which their flanks can be shaped automatically and precisely. He applies for a patent, first in Switzerland, then in Great Britain [100]. His plan is to develop his idea further for practical applications; for that purpose, he even has financial support in the form of 3000 Swiss Francs from the Benno Rieter Foundation in Winterthur. He begins work on the project, but cannot complete it, and finally gives back the main portion of the grant.

In spite of his poor health, he takes a great interest in world affairs, reads several newspapers and follows the progress of science meticulously in the literature. His daughter Elsbeth recalls that his correspondence with associates from the Mathematical Society continued unabated. It is a special pleasure for him to receive one of the occasional letters from Albert Einstein. Grossmann at first writes by hand with a pen and black ink, even though his right side is more severely affected by the paralysis symptoms. Later, he types on a typewriter using his left hand. He even administers school-leaving examinations at home; in his son's room, writing materials are kept ready for the candidates on a large table.

The costs for medical care, medications and his stays at health spas are a drain on his pocketbook as an early retiree, who never took up a health insurance policy and has to pay all the costs out of his own funds. He indeed has a good pension, but it is far from sufficient to allow the family to continue living in their accustomed style. The mathematician never saved money during his good years, when he received a stately professor's salary. His slim reserves have to be drawn upon and are quickly exhausted. Nevertheless, father Marcel doesn't let himself deprive his son Marcel of support for a three months' stay in Paris as a student, and his daughter Elsbeth can finish off her high school studies. Like his father before him, he limits his own expenditures rather than saving money on the education of his children. There is seldom meat on the table; Anna does her best to cook delicious meals even with low-cost ingredients. Since the family can no longer afford to hire a maid, she takes care of all the housework herself.

But in spite of great discipline and making use of all the possibilities for saving money, the family patriarch Jules has to be asked for money more and more often, in order to finance the cures, medical treatments, doctors' bills and medications. The resulting conversations often lead to arguments, putting a strain on father-son relations. Financial worries overshadow their family life to an increasing extent. They can only rarely invite guests, but their contacts to a few loyal friends are maintained. There is a hearty friendship especially between the Grossmanns and the family of Rudolf Fueter, professor at the University of Zürich and co-founder of the Swiss Mathematical Society. With the Fueters, Marcel is never ashamed of his motoric problems; they have a daughter who is severely handicapped and have become experts in that area. The Grossmanns also maintain friendly relations with Mileva Einstein and her sons, although they are never very close, and Albert comes by to visit whenever he is in Zurich.

In 1928, Marcel Grossmann consults the pharmacologist Alfred Jacquet at his private clinic '*La Charmille*' in Riehen, and at first obtains good results from his therapy programme, which was developed for the treatment of cardiac and circulatory problems. Unfortunately, his symptoms return after a few weeks. A year later, he tries a cure in Degersheim, and two years later in Herisau. Conventional medicine, alternative healing methods, autosuggestion, cures at the seaside and on land: Anna and he try everything that is within their increasingly limited means. But the illness continues its advance, and the mood in the *Holderstraße* becomes darker and darker.

13

Rosenau

How did life continue on the other side of the lake, after Jules and Mimi Grossmann had moved to Thalwil? The two of them made themselves at home in the *Villa Rosenau* and kept an open, hospitable house there. Amélie, the daughter from Jules' first marriage, moves to Thalwil after working in Mecklenburg-Schwerin and again begins to study at the arts and crafts school, but once again drops out. It speaks for the tradition-conscious Jules that he doesn't press his indecisive daughter to marry, who at 29 has reached the upper age limit for marriage according to the notions of that time, but instead supports her practising a profession. He offers the young woman a position in the family enterprise, for which she would be well suited with her talent for languages. Amélie declines, but at least she helps her step-mother Henriette whilst she is living in the *Villa Rosenau*.

© Springer International Publishing AG, part of Springer Nature 2018
C. Graf-Grossmann, *Marcel Grossmann*, Springer Biographies,
https://doi.org/10.1007/978-3-319-90077-3_13

Photo Eugen, Amélie, Jules and Henriette Grossmann, Thalwil (from left to right, 1899)

The tall, slender oldest daughter turns out to be a modern, independently-thinking woman with a strong will, who resists bending to the opinions of others and defends her own, at times gruffly. She thus often rubs people the wrong way, and she argues frequently with her half-brothers. In particular, Amélie and Eugen occasionally have serious disagreements, usually over money matters. Most of the time, however, the discussions are peaceful. Grandson Marcel Jr. later recalls that in *Rosenau*, there was often *"political talk"* when the father and his sons came together. They debate over social and economic questions, over the rights and duties of workers and employers; Father Marcel and Grandfather Jules hold a rather liberal-conservative position, whilst Uncle Eugen is attracted by socialism, although he is not a member of any political party. The youngest brother, Eduard, sits in the meantime on the Communal Council of Thalwil, which gives him a different view of community affairs. The boy Marcel doesn't yet understand in detail just what is being debated, but he recognises that they are often not in agreement. Jules sometimes vociferates against the aristocracy; at such times, the French Republican arises in him. In daily life, however, he was, according to his grandson, *"the opposite of a revolutionary, namely the embodiment of a practical merchant and entrepreneur who was interested above all in personal property"*. One could add that Jules generously overlooks the fact that his Mimi comes from a formerly aristocratic family, and that her brother Hans married Dolly von Wattenwyl, a representative of the Bernese patricians.

Jules Grossmann operates his business in the cotton factory with modest but continuing success. The First World War initially causes him problems, but soon gives rise to a powerful boost in sales: Following the declaration of war on August 1st, 1914, the factory operations are shut down until the fall, because son Edi, who has in the meantime become indispensable, as well as the foreman and the factory workers, have all been mobilised and are on active military duty. In the fall, however, *Patron* Jules can start up operations once again; he has received a large order from France for cotton bandages. He is evidently able to round up a sufficient workforce; even his wife works along with the others. In 1915, he shuts down the production of bandages and rapidly increases production of cotton padding; the Swiss army is outfitting its personnel with the new field-grey uniforms of the Ordinance 1914/17. Jules has to produce enormous amounts of cotton padding for shoulder pads within a record time. The lack of German competition and the good economic situation within Switzerland give the business a new impetus from 1915 on; profits increase to a multiple of their previous levels. Now, he can even afford his own secretary, Ms. Alice Abegg.

Financially, the Grossmanns are doing much better after 1914. Politically, the family is united on the French side. Paradoxically, however, the fact that the family business has finally become profitable has led to differences of opinion and to tensions. Much to the chagrin of their mother, the two elder sons have the impression that Father Jules is giving preference to their siblings, and passing on to them for the most part the benefits of his now more freely-flowing monetary funds. Eugen Grossmann, in his memoirs, even goes so far as to declare himself and Marcel to be *"war victims"*. This seems somewhat strange, given that their parents have allowed each of the children to have a carefully chosen and appropriate education; none of them had to make sacrifices for financial reasons.

Jules Grossmann is evidently of the opinion that his two older sons can stand on their own feet now. And they can—Eugen has by now also embarked on his career—but Marcel's salary at the ETH is apparently barely enough to meet the needs of his growing family. Eugen describes how Marcel's wife and daughter are beset by illnesses which are related to undernourishment. He notes bitterly that their parents did not even *"take the warning of these alarming symptoms seriously, but instead continued to leave the 'war victims' to deal with their fates alone"*. This results in an unpleasant disagreement, which disturbs relations between the two elder sons and their parents for a long period beginning around 1916. Later, the situation again normalises, but the breach of trust is there, and still later, after the parents' deaths, it will lead to an open rift in the family.

Photos *Above*: *Rosenau*, Thalwil | *Below, behind*: Jules, Anna, Mimi, Amélie and Hedwig Grossmann, Anna Keller; *in front*: Henriette and Elsbeth Grossmann (from left to right, 1924)

Jules Grossmann fails to mention these tensions in his memoirs. Undaunted, he continues to work, following his motto, "*les affaires avant tout*", although he is in the meantime 72 years old. In the middle of the War, he goes on a business trip via Pontarlier to Paris. A year later, his firm participates in the Trade Fair at Lyon; he sets up the stand himself, as always, and is present the whole time. In 1917, he briefly considers increasing the production capacity of his cotton factory in Thalwil by adding some carding engines, but then he decides not to carry out the expansion—his experiences in Hungary have not been forgotten. Near the end of his business career, he finally manages to accrue some financial reserves, which must have comforted him considerably. Very light-heartedly, the family celebrates the 80th birthday of its patriarch on March 2nd, 1923. Everyone comes—children, grandchildren, members of the Keller and the Lichtenhahn families. The guests in Zug have no idea that this will be the last celebration at which all the family members can be present.

The following year brings a roller-coaster ride of emotions for the old couple; it begins happily with a family get-together for New Years in the "*Rebstock*" guesthouse in Thalwil, at which the senior of the family announces that the direction of the firm will henceforth be laid in the hands of his son Eduard. In the spring, the two elderly spouses spend a happy time in Weggis on the *Vierwaldstättersee*, and then stay for a few days in Montreux. But in the second half of the year, Mimi's health begins to worsen. Jules and Henriette indeed travel to Montreux again in the fall, but the wife suffers an attack of cardiac insufficiency whilst there. The old lady recovers, however, and notwithstanding her growing weakness, she begins preparations for Christmas. Although she has a cold, she won't be deterred from making handmade presents for her children, grandchildren, friends and godchildren, and personally doing her shopping in Zurich. She also seems to follow the *Methode Coué*, and rasps untiringly that she isn't at all hoarse… One evening, Mimi comes home, dead tired. Three days later she has another attack of cardiac insufficiency. Frightened, she calls out, "*I am empty!*" and sinks down onto the sofa. The doctor who is quickly called makes a serious diagnosis: weakness of the heart muscle. At Christmas, the old lady however feels better and is able to attend the little family ceremony. Some months pass, Mimi takes a bit more care of herself, and she is looked after several times a week by two of the Sisters of the Order of the Bethany Deaconry.

On June 19th, a strange thing occurs: The invalid suddenly stops reacting, and towards evening, her heart nearly ceases to beat; her body cools off. The doctors see death approaching, and there is talk of preparing the funeral. Jules Grossmann sits beside the bed of his companion, holds her cold hand in his and all of a sudden feels that it is slowly again becoming warmer! Gradually, life returns to the still figure, she awakens and whispers with astonishment, *"My saviour has not yet come."* The doctors have no explanation for this phenomenon. Her husband is ecstatic, and believes that his Mimi is on the road to recovery. But on July 5th, she suddenly suffers strong pains in her feet—her cardiac insufficiency, and as well the time spent lying in bed, have led to a lack of blood circulation in her lower extremities, and the tissues in her feet are suffering from a creeping necrosis. Her symptoms have disappeared again the next day, and when her granddaughter Elsbeth comes to visit, Henriette is able to contemplate her for a long time, holding hands with the young girl. Days later, the fearful pains have returned, and the situation is now clear: the lack of blood circulation has led to gangrene. Today, such a condition, at that time called 'senile gangrene', could be treated; but for Mimi, it is a death sentence. On July 9th, she wakes up, sits up in bed, crying, and takes her farewell from Jules. Thereafter, she doesn't wake up again, and on July 11th, 1925, she quietly stops breathing. Her husband only now feels his *"enormous misfortune. Everything that at the time could be done to save her was tried; two Bethany Sisters cared for her faithfully."* Jules Grossmann has become a widower for the second time, and he has to let his Mimi go. He has only one hope now, *"that my soul will come to the same place where I know that the soul of my dear spouse is resting"*.

Adding to his desperation over the painful loss are practical problems: What is to be done with the big house? Various possibilities are considered. The old man admits himself that he hardly takes part in the discussions; he can barely formulate any clear thoughts. Amélie makes the audacious offer of looking for an apartment on Lake Geneva for herself and her father, and finding work as a translator. For the time being, a cousin comes from Münster to keep house for the widower. He also has a household employee, Betti, whom he however has to dismiss because of her *"pathological breaches of trust"*. The lack of a housewife who can look after things is noticeable all around the place. During this difficult time, Jules Grossmann becomes gratefully aware of just how many people in Thalwil had loved Mimi; from all directions, offers of help and sympathy arrive. In 1927, the children discuss the question of whether their father should move into a retirement home. Then a lucky opportunity arrives: Ms. Seline Gaehler, a relative of their old friends from Budapest, the Gaehler-Haggenmacher family, offers to move to Thalwil and serve as housekeeper for Jules. The sons discuss this possibility and all of them are gratefully in favour of accepting it. Seline Gaehler remains in the *Rosenau* until the end of her life.

Photo Henriette Grossmann (1923)

Jules Grossmann is still remarkably mentally alert and capable. At 86, he complains that his near vision has gotten very poor and that he has difficulties reading print, typewritten text and handwriting. But his distant vision has gotten better! The old man goes about without glasses in his house and garden, and he is visibly proud of that. He is filled with a quiet optimism, and enjoys observing the daily lives of his sons and the maturing of his grandchildren. The serious illness of his eldest son is indeed registered with concern, but he hardly mentions it in his memoirs. He greatly enjoys the engagement of his grandson Marcel Jr. to Hanna Appenzeller, and their marriage on October 17th, 1931. The lovely young woman is pleasing to the patriarch, and he reports happily on the wedding ceremony in the little Witikon Church, on the flowery decorations and the bridesmaids, the pastor's serious sermon and the fine wedding feast held at the Guildhall of the Carpenters in Zurich.

In 1923, Jules dictates his memoirs to his granddaughter Elsbeth, who is serving as his private secretary. They fill 252 pages, painstakingly typed on high-quality handmade paper by Elsbeth and bound in solid leather by a bookbinder. Although his vision is no longer acute, the senior takes it upon himself to read through the book meticulously and add corrections and additions in his fine handwriting. On March 2nd, 1933, a roaring celebration in honour of Jules' 90th birthday is held. In the elegant Guildhouse of the Carpenters, family and friends gather in evening wear; Anna comes with Marcel Jr., his wife Hanna and Elsbeth; her husband can no longer get out of bed.

Jules Grossmann has in the meantime gone completely blind. He makes a speech, thanks all those present movingly for their thoughtfulness and looks back over his life. The old gentleman recalls Mimi's role in his life and admonishes his family to live their lives to the best of their knowledge and beliefs. The disorders of his age are becoming more and more noticeable; he often stays in bed all day, and speaking costs him a considerable effort. He is calm and confident, at peace with himself, and often speaks of his "*dear Amanda and dear Henriette*". On the Swiss national holiday, August 1st, 1934, Jules dies, aged 91, following a stroke. Marcel Jr., who in the meantime is working at the *Schweizerische Rückversicherungs-Gesellschaft* [Swiss Reinsurance Corporation] as a macroeconomist, notes in his diary, "*An epoch has passed. In the evening to Thalwil, he is lying there in his bed, a handsome, striking face, full of energy and willpower. That was a man!*" [101] Peter Marti, the son of Mimi's sister Ernestine Marti, who like his father has become a pastor, honours the patriarch at his funeral two days later with warm words: "*That was the dear departed: spirited, ardent in his sympathies and antipathies, as an employer exacting and benevolent at the same time, as father of a family tender and kindhearted, at his very deepest with a childlike humility* [102]."

Photo Jules Grossmann (1933)

14

Friends

As was already the case on his father's 90th birthday, Marcel Grossmann could not attend the funeral ceremonies for his father, and he had to accept the painful realisation that life goes on without him, in spite of all of the sympathy of others. A certain bitterness is understandable, and it is exacerbated by his feeling of dependence. The mathematician spends all of his days at home, working sometimes at his desk or, more and more frequently, in bed, propped up on thick pillows. Since his handwriting has become almost illegible, he balances the typewriter on his knees and types letters laboriously, mostly with his left hand.

The fact that he can no longer leave the house and receives practically no guests has one advantage—the distressing problem of his wardrobe no longer plagues him. His daughter Elsbeth recalls all too well the exhausting excursions that her parents undertook to the city to buy clothing and shoes. Even climbing onto the trolley cars requires a strenuous effort on the part of her father, and he is frequently stared at by other passengers and shop personnel because of his clumsy gait and his unclear speech. He sometimes cries after such humiliating excursions, and is more and more often depressed. If he didn't have his contacts with acquaintances and friends via correspondence, his desperation would probably have completely overcome him. Even so, Grossmann's melancholy mood casts a shadow on his relations to his family and his environment. Marcel Grossmann is not a cooperative patient; he is chubby and ought to lose weight so that he could be more easily cared for. But who would be surprised: The gourmet objects vehemently to the efforts of his wife to hold him to a diet; he simply likes goulash better than

C. Graf-Grossmann, *Marcel Grossmann*, Springer Biographies,
https://doi.org/10.1007/978-3-319-90077-3_14

vegetables, and prefers the favourite pastries from his childhood in Hungary to the less rich dishes that Anna offers him. This is certainly one reason why the mood in the household is often peevish. In addition, the two sisters, Anna and Hedwig, argue increasingly often; their styles of life and opinions are just too disparate. What does Anna feel when Eugen and Hedwig return one more time, cheerful and suntanned, from Capri, Portofino or Venice? When Hedwig models her new dresses and reports on glamourous evening invitations? As a rule, the older sister controls herself and swallows her frustration, but often enough she loses her patience.

Eugen tries to mediate between the two women, because the tensions under the same roof are hard for him to bear. The younger professor is close to his older brother, often sits by his bed and recounts amusing episodes from work at the University, trying to distract him and cheer him up. There isn't a great deal of positive news to report; in October 1929, the worldwide economy begins to crumble after the stock-market crash on Wall Street, and Europe is still suffering from the aftereffects of the World War. Eugen Grossmann, who as an economist occasionally advises the Zurich Governing Council and the Swiss Federal Department of Finance in Bern, observes these developments with concern. And the progressive illness of his brother depresses him more than he is willing to admit even to himself. In the end, he cannot bear life in the "*house of misfortune*" in the *Holderstraße* any longer. In 1931, Eugen rather suddenly buys a house in the *Stüssistraße* and moves there with his family. Anna is relieved on the one hand that daily life again becomes calmer and more peaceful, but she misses her energetic brother-in-law, with his competence and good advice.

Their son Marcel Jr. has also led an independent life for some time now; he lives with his Hanna in Witikon, above Zurich. Elsbeth, in contrast, remains at home, hardly leaves the side of her overworked mother, accompanies her on her rare visits to family or to the theatre, and gives her moral support. When, later on, Marcel can no longer get out of bed at all, Anna carries almost all of the burden, although the district nurse does help occasionally.

Anna Grossmann not only takes care of the household, but also of the family's business affairs and their finances. Her husband seems to have little interest in the household worries; he lives increasingly in his own world of ideas, whilst his strength gradually dwindles. Elsbeth's main task is to keep an eye on her father. When he is sitting at his desk and then struggles to stand up, he sometimes falls over. If he is working in bed, she brings him books, paper and writing materials—and keeps him from setting the house on fire! The patient refuses to give up smoking his pipe, but he can no longer bend over. After a precarious incident with a burning match in the waste-paper basket, Elsbeth keeps watch over him, sits beside him, working on exercises or reading. Or she dances and practises gymnastics in the next room, since the daughter of the family is an enthusiastic ballet dancer (she later becomes a teacher of gymnastics and rhythmics).

In these gloomy years, it is a blessing that the mathematician has kept up his contacts to colleagues in the Swiss Mathematical Society and the Society for Natural Science Research. His dialogue with Albert Einstein is also revived during the 1930's. Einstein is still living in Berlin, and has become world famous since the confirmation of General Relativity in 1919 and his winning the Nobel prize. He is also politically very active, and reaps both support and criticism, the latter with increasingly racist tendencies. But Einstein defies the growing antisemitism. Marcel follows Albert's publications on General Relativity theory and its implications and further development. Einstein is searching for a "universal formula" which would describe all the forces of Nature in a unified fashion, and would include space, time, and matter. It should also unify relativity theory with quantum physics. An extremely ambitious project, which is still an unsolved problem today.

Einstein is attempting to unify gravitation with the other [at the time known] fundamental force, electromagnetism. In 1928, he publishes two articles on this topic in the Proceedings of the Prussian Academy of Sciences [*Sitzungsberichte der Preussischen Akademie der Wissenschaften*], calling his new approach "distant parallelism" [or "teleparallelism"]. He has discovered that the gravitational field can be characterised not only—as in General Relativity—by the curvature of spacetime, but rather that there is an alternative mathematical formulation which permits a further generalisation, in which gravitation is described in terms of the "torsion", i.e. a kind of twisting or vorticity of spacetime [103].[1] Einstein is searching for a revision or extension of the theory which will yield a "unified field theory", and continues working towards that goal—unsuccessfully—until the end of his life.

In General Relativity theory, the field equations are obtained by computing the Christoffel symbols and the Riemannian curvature tensor. In distant parallelism, in contrast, it is the torsion which plays the role of a gravitational force. Simplifying, one could say that the two approaches are like the two sides of the same coin, the 'coin' of the formulation of gravitation [104].

---

[1]The mathematical basis and some modern applications of distant parallelism, as well as a few descriptive analogies, can be found in the *Wikipedia* article on 'Teleparallelism'; see https://en.wikipedia.org/wiki/Teleparallelism.

Here, Einstein ventures far into the field of mathematics and into quantum theory. In 1929 and 1930, he publishes nine further articles on relativity theory with torsion, and carries on a lively correspondence with the French mathematician Elie Cartan. His friend Marcel also feels himself to be directly addressed. He requests that Einstein send him reprints of his latest publications, and receives them immediately. Marcel is extremely critical in regard to Albert's new theory, but he expresses this in a relatively jocular manner. On November 23rd, 1930, he responds to a letter that Albert had written to him on the noble letterhead writing paper of the Prussian Academy:

Dear Einstein, For a long time I have brooded over the problem of finding a letterhead which would be even halfway worthy of the one that you used. In vain. I spent less time thinking about the content of your letter. The approach is simply false, for one cannot construct a Riemannian manifold from 4-vector fields. [...] I will soon write up what I have to say against Levi-Civita, Cartan, you, Weitzenböck etc. But I do not wish to cause you any kind of difficulties, since I know that there are swarms of extremely stupid people out there.

So what suits you better: Should I send you my manuscript (about 10 printed pages) in early December to be presented to the Berlin Academy & published in their Proceedings; or should I publish it here in Switzerland? Please reply soon. Hearty greetings, Yours sincerely, G.

P.S. In any case, I am not in need of protection & can still defend myself against arguments. [105]

From Herisau, where he is taking another health cure, Marcel writes in April 1931 to Einstein in a short, whimsical letter:

Dear Einstein, Enclosed are discussions which demonstrate that the fundamentals of your current theory are mathematically naive. I can give you no advice as to your answer; only the remark that I am in no need of protection, i.e. no testimonium paupertatis. I am well along towards completely recovering all of my faculties. It was, as I have presumed for years, a disturbance of the blood circulation. But still I say to you, freely quoting Heine: 'Don't disgrace yourself, oh Ferdinand / And let yourself be converted / Later on, when your mind is again free / It will provide you with something better.' Yours, G. [106]

Marcel Grossmann remains insistent, but Albert Einstein does not allow himself to be converted. On June 9th, 1931, he writes his friend a long, patient and friendly letter:

Dear Grossmann, I have read your treatise on 'distant parallelism' with care, and I can see from it that my two letters were not capable of convincing you of the <u>logical</u> correctness of my new theory of space. I have also now understood just where you see the difficulties. A public dispute does not appear to me to be expedient, since the objections that you have raised do not seem thus far to have been felt elsewhere. That is not to say that I disapprove of the fact that you yourself have chosen to publish your opinions. I simply wish to justify to you why I answer your discussion only privatim.

This is followed by over three pages of detailed scientific arguments. He closes on a conciliatory note: *"Dear Grossmann! Forgive me for the length of this letter. But I think that you will have fun with it. Permit me to greet you and your lovely wife heartily, Yours sincerely, A.E"* [107].

'Fun' is a word that no longer occurs frequently in Marcel Grossmann's vocabulary. All the more must this last letter from Albert—in spite of the polite but clear admonition—have distracted him and led him to new thoughts. He rouses himself sufficiently to reply, although one can see from his typed letter how much effort it had cost him to write it. His letter of June 23rd, 1931, was sent by registered mail, so that he could be certain that it had reached its destination.

Dear Einstein, Many thanks for your detailed letter; although it only shows that we simply do not understand each other. But your letter was all the nicer, for you are convinced that I am in error. I, however, am sure of my position.

I must say that the critical part of your letter is somewhat meagre. You consider only two points. [...] One cannot just throw a correct mathematical theorem out the window and say e.g.: 'Tomorrow morning, from 7 o'clock on, the Erlanger Programme is no longer valid – because it contradicts a new theory.' Then the new theory is simply false, I would say. Naturally, it is all the same to me, if you believe that you are in the right, not to answer publicly. But many mathematicians and physicists agree with me. I retain for myself all the freedom to speak out.

So X on ahead with Mayer, who is, along with Duschek, also in error in this matter. Your pickles will come back up, as you like to say, & then we will understand each other again.

Come to visit us when you are passing through. I am slowly getting better, since I have understood the nature of my problems & am fighting them by eating a proper diet.

With warm greetings, from my wife as well,

Yours sincerely, your unpredictable but faithful friend, M.G. [108]

These two letters from April and June demonstrate that whilst Marcel Grossmann was by now arguing rather obstinately, he was still mentally fit and thoroughly spunky. In 1932, Einstein will abandon his *teleparallelism* approach to the "universal formula", but not for the reasons advanced by Marcel Grossmann. However, the geometrical theory of distant parallelism, which was also independently discovered and developed by the French mathematician Élie Cartan, and is now known as the Einstein-Cartan theory, is still of interest to theoretical physics today.

The confidence with which the mathematician describes his physical state is however astonishing. He believes himself to be on the way to a full recovery of his faculties; he is improving. Is he once again applying the Coué method and practising positive thinking? It is certain that he was clutching at every straw as far as the explanation of his illness was concerned. Apparently, during the cure at Herisau, they had suggested to him that he was suffering from a harmless disturbance of the blood circulation, and that had restored his courage. Although he was otherwise a critical person, he takes each ever-so-improbable explanation of his illness to be the real thing. Anna supports this, much to the annoyance of her brother-in-law Eugen, who is an adherent of conventional medicine. In his eyes, she is searching for help from largely dubious alternative healing practices.

The facts speak clearly for themselves. Whether or not Marcel wants to accept it, his multiple sclerosis is running its course. Ahead of him are five years which can only be described as a nightmare with brief periods of relief. In the spring of 1932, he suffers an attack of pneumonia, with 104° fever, and for the first time he is forced to go to hospital, to the *Paracelsus Clinic* in the *Seefeld* District of Zurich. Just why he was brought there has not been recorded. The location no doubt played a role; the hospital is only 20 min by foot from the *Holderstraße*. Furthermore, its reputation and that of its doctors and the Menzinger Order of nurses who work there is excellent. Marcel Grossmann recovers, although before the advent of antibiotics, there was little that could be done for pneumonia patients. There were experiments with various sera, but mostly, only the symptoms could be mitigated, pain reduced and hoped that the body's own immune system would overcome the illness. Indeed, the mathematician musters all his energies and recuperates, and he can return home from the clinic.

Grossmann had liked the clinic; he got along well with the Menzinger nuns, who are efficient and kind. Although a Protestant, during his time as president of the Swiss Federal Commission on Higher School Leaving Certificates, he had met up with the world of the Catholic cloisters and their schools. He values their humanism and the seriousness with which the priests, in particular the Jesuits, had for centuries instructed their students in the liberal arts, including geometry, arithmetic and astronomy. That he experienced a feeling of well-being in the clinic, in spite of his medical problems, is a godsend for Anna. Her husband can now spend occasional holidays there, so that she has some much-needed relief from her duties.

The *"prettiest girl of Thalwil"* is by now 50 years old. Her youthful freshness has vanished, and the shy fiancée has become a self-assured, dignified professor's wife, who strides straight ahead through life and wears her hair long and knotted up into a bun, although women's fashion in the meantime prefers short, pert hairstyles. On June 18th, 1932, her birthday is celebrated at home, cheerfully and with a good meal. Her troubles are forgotten for a few hours, but soon after, they come back with a vengeance. Anna calls her son to a crisis meeting in the *Holderstraße*: The baker who had been the renter of the ground floor shop has given notice. Anna Grossmann considers selling the house; the load is simply too great to carry, and the apartment without a lift is not suited to caring for a chronically ill person. Marcel Jr. promises to help, and soon afterwards, he visits Director Gottlieb Duttweiler in the offices of the *Migros* company on the Limmat Square. The 45-year-old entrepreneur and politician receives the young man amicably and considers setting up a bakery in the *Holderstraße*. A year later, he carries out this project, but unfortunately, the baker whom he commissions soon terminates the contract with the excuse that the oven there consumes too much fuel…

Anna Grossmann is despondent; she needs the rent payments badly, and her son is annoyed by this failure. In January 1934, a serious discussion on the question of the care of Father Marcel takes place in the Villa *Rosenau* in Thalwil, a regular family council. What thoughts might have gone through Anna's mind during the train ride back to Zurich? Can she gather her strength again with the help of the solidarity of the family? It is present not only in the Grossmann family; Anna's mother, who has been living in a small apartment in the *Gotthardstraße* in Thalwil since the death of her husband Eduard Keller, helps out her daughter frequently with guidance and resources, and some banknotes now and then. And the fulfilling life of the two children, Marcel and Elsbeth, also offers some consolation; on July 12th, 1934, Marcel's and Anna's first grandchild, Urs, is born.

The situation does not become easier after Jules' death on August 1st, since now, it must be decided what is to be done with the *Rosenau* house. The factory is taken care of; it has been under Eduard's management for the past two years. But the stately and meanwhile quite valuable house and its furnishings are the cause of a proper quarrel. The whole family gathers numerous times for meetings at the Villa *Rosenau*. According to Marcel Jr.,

Amélie proves to be *"enormously egoistic, whilst Mama and I are more or less the scapegoats for* tout le monde". The Villa *Rosenau* must be sold; it already seems like a *"body without a soul, a house demolished, a destroyed family; gruesome"*, the young man records despondently.

The family quarrel escalates in January of 1935; Jules' oldest daughter causes difficulties and finally walks out. Eugen rants and fumes, he curses Amélie, whilst his wife Hedwig and daughter Henriette sit pale with fright beside him. The next day, in yet another meeting, the sale of the Villa *Rosenau* is sealed. The invalid mathematician seems not to have taken a stand on the controversy surrounding his parents' house; whether from insufficient strength or a lack interest can no longer be reconstructed. In December 1935, Marcel Jr. writes forebodingly in his diary, *"1936—already another new year, what may it bring?"*.

It is principally the political situation which provides fuel for conversations and a reason for concern. In Germany, Adolf Hitler has been chancellor for the past three years and has lost no time in giving that country back the power which it in his opinion deserves. The 'Third Reich' leaves the League of Nations, and in March 1936, events chase each other in rapid succession: Hitler calls up the *Reichstag*, terminates the Locarno Pact, in which the German Empire, Belgium, France, Italy and Great Britain had guaranteed the mutual respect of their current borders, and sends German troops on the same day into the Rhineland to abrogate the demilitarised zone there. He thereby puts the League of Nations, in which many people had placed great hopes, in a nearly insoluble situation. The newspaper *"Neue Zürcher Zeitung"* reprinted the content of the meetings of the organisation almost verbatim, and reading them, the ups and downs of hope for a diplomatic solution become apparent with an oppressive clarity.

Not only on the political front are storm clouds gathering; in Switzerland, the weather is also acting up. After an initial warm period, by Easter a thick covering of snow is lying on the flowering trees. It abruptly melts, making way for midsummer temperatures. In June, when Marcel Jr. reports to Lucerne for his military reserve continuing training, the country is suffering under a leaden heat. In July, when Urs' second birthday is being celebrated, perspiration is flowing freely. How does Anna manage to keep her immobile husband cooled, prevent him from getting bedsores, and maintain a certain dignity for the patient? She succeeds by mustering all her forces so as not to suffer a breakdown herself. Every walk to the mailbox is like running the gauntlet, since bills flutter with an unnerving regularity into her house and must be paid. Even in these difficult times, when money is lacking in every nook and cranny and she is sometimes totally exhausted, Anna Grossmann manages to contribute 200 Francs for a health cure for her daughter-in-law, who is suffering from persistent neck pain.

In early September, Marcel Jr. becomes ill with double pneumonia, and shortly thereafter, similar symptoms appear in his father. A unique constellation: These two men, who to be sure bear the same name but have never been particularly close, suffer from the same illness at the same time. The mathematician, already weakened, will not recover. His son is lying severely ill and helpless in bed when the news arrives that his father has passed away on Monday, September 7th, 1936 at 21:15 h in the Paracelsus Hospital. He is told that his father was unconscious and did not suffer. Whether this was in fact true, or whether the news was intentionally moderated for his protection, will never be known. Marcel cannot even attend the funeral; he is totally weakened after surviving pneumonia. His father's urn will be buried in his absence at the Enzenbühl Cemetery in the *Forchstraße*, in the Zurich-*Hirslanden* District.

Louis Kollros, Grossmann's fellow student and colleague, writes an epitaph for the Proceedings of the Swiss Society for Natural Science Research. Two additional epitaphs were written by the mathematician Walter Saxer. He had studied under Grossmann, became his assistant and, following Marcel's retirement, was his successor as professor of descriptive geometry, and later Rector of the ETH. For the Proceedings of the Zurich Society of Natural Scientists, Saxer writes:

> In the night from the 7th to the 8th of September, 1936, Prof. Dr. Marcel Grossmann was released from his many years of severe suffering. His numerous former colleagues and students know that death was a deliverance for him, and they recalled in this hour of farewell once more, and with particular thankfulness, their former teacher, the well-known researcher, the notable human being. Marcel Grossmann belongs among those people whose life's work can never be described in words; one had to have experienced it in its versatility in order to fully appreciate it. [...]. His first mathematical works dealt with non-Euclidean geometry. Even as a young Canton School teacher, he developed rather nice planimetric constructions for non-Euclidean geometry, and they were praised by none less than D. Hilbert. However, the most important and most famous mathematical achievement of Grossmann's was to provide Einstein with the first mathematical concepts for his theory of gravitation – their classic joint publication appeared in this journal in the year 1913. Grossmann's contribution to this publication is often underestimated; in fact, he did Einstein an important service by working together with him as a mathematician. Grossmann, for his part, was very happy with this joint research effort, and only the outbreak of the War caused the abrupt interruption of a

hopefully-begun period of research. Other mathematicians have recognised the correctness of Grossmann's ideas and have fundamentally extended and elucidated them in all directions. [...] Grossmann's lectures on descriptive geometry, which were attended by all the German-speaking engineering students of the ETH, demonstrated exemplary clarity and eloquence and always aimed at building as many bridges as possible from geometry to technology. [...] Along with his classes in descriptive geometry, he taught mainly synthetic geometry in the Department of Mathematics and Physics. His influence here was of decisive importance. During the first years of his tenure at the ETH, various dissertations were completed on non-Euclidean geometry under his guidance. Later, it was less his scientific influence and more his general human qualities which made him the real leader of the Department. As its chairman, he set up various new and quite valuable institutions. For example, during that time, pedagogical-didactic lectures for the teaching of mathematics were introduced.

Grossmann's work as an organiser radiated its effects in all directions. He served the ETH on many committees and commissions – for 6 years, as a particularly valued chairman, he led the large Department of Mechanical Engineering. He worked tirelessly to organise the Swiss mathematicians and to ensure that their work received its just resonance within and outside the country. [...] As a leading member of the Federal Commission on Higher School Leaving Certificates, he fought courageously in the great debate over school reform, tirelessly pointing out the national goals. He threw his last and best energies into that debate. It was no doubt a serious disappointment for him as a sick man to stand by and see, without being able to intervene, just how little finally resulted from the whole reform process. [...] As a clear and intelligent man of action, he always put all of his energies into reaching the goals that he had determined to be right and proper. He expected discipline and participation in the great front of human efforts in the service of imperishable intellectual values from everyone else, and most especially from himself. His straightforward and truthful nature, his generosity and genuine humanity, and his works will not be forgotten. [109]

Saxer also wrote a long epitaph for the "*Neue Zürcher Zeitung*". In two columns, he dedicates similar words to the memory of Marcel Grossmann as in the Quarterly of the Society of Natural Scientists, but adapted to the readership of the "*NZZ*":

There are men for whom relaxing is a torment; Marcel Grossmann belonged to that group, and a truly tragic fate condemned this man of great vitality and intelligence, always ready to act and to accept responsibility, to a lingering illness. All of his many former colleagues and students have noted with deep emotion that in the night from the 7th to the 8th of September, death came to release him from his sufferings. He had long since taken his leave from us and lived *a secluded life within his family.*" Among Marcel Grossmann's accomplishments at the ETH, Saxer mentions his function as a bridge-builder between the different departments: "*It speaks for all those involved that the large Department of Mechanical Engineering called him, a mathematician, to be its chairman for six years, although it had its own highly qualified members such as Stodola [...] among its ranks. In that capacity, Grossmann was not always an easygoing colleague. He could pursue a goal that he had recognised to be worthy with great energy, and sometimes very clearly pointed out the exaggerated individualism of teachers at all levels. However, he always applied his high standards of objectivity to himself as well, and he knew his own limits quite precisely. [...] During the War, Marcel Grossmann proved that he was not only a good Swiss citizen, but also a good European. He was aware of the much greater suffering in the warring nations and made an effort to alleviate the misfortune of students who were taken as prisoners of war, through an extensive relief organisation. He made many intellectual and material sacrifices for that organisation even after the end of the War. [...] Now, Marcel Grossmann has passed on. He made great efforts to lead a life compatible with the definition of a scholar in the sense of Fichte, by filling the position allocated precisely to him with dignity* [110].*

From all over the world, letters of condolence pour into Anna Grossmann. Hans Albert Einstein [Albert's son] writes a cordial card to her on September 11th, 1936. Mileva Einstein also bestirs herself to write a few lines. On September 20th, 1936, the widow receives a letter from Princeton, which she guards jealously for the rest of her life:

Dear Mrs. Grossmann,

Yesterday, under a mountain of unopened letters, I found an envelope edged in black and opened it. I saw there that my old, dear friend Grossmann had been relieved of his sufferings – a grim fate after his hopeful and productive early years. And you suffered along with him and bore up with a wonderful vital strength. I must say that you belong among the few women of my generation for whom I feel a truly genuine admiration and respect. Bearing up and giving one's life without hope, that is the hardest thing, which a man practically never manages to do.

Memories of our shared time as students rise up in me – he, a model student; I, disorderly and dreamy. He, closely associated with the teachers and readily comprehending; I, off the main track and dissatisfied, not well liked. But we were good friends, and our conversations over an iced coffee at the Metropol every few weeks belong among my most pleasant memories. Then, at the end of our studies – I was suddenly left quite alone, standing at a loss before life. He however did not abandon me, and through his help (and his father's), I came a couple of years later to Haller at the Patent Office. It was a kind of lifesaver, without which I would indeed not have died, but would have atrophied intellectually.

A decade later, our mutual, feverish research work on the formalism of General Relativity… It remained unfinished, because I went off to Berlin, where I continued working alone. Then his illness soon began; during the student days of my son Albert, the first symptoms became apparent. I have thought of him often and with pain, but we saw each other only rarely, when I was there on visits.

I would never have dreamed that it would stretch out so cruelly long, although I am familiar with that disease from a friend in Berlin. Now, he has passed away, but not before I have become an old man, after having experienced the most diverse strokes of fate – uninvolved within my inner self (?) – and now perhaps a few years of quiet existence still remain for me.

But one thing is still beautiful: We were and remained friends throughout life.

To you, though, I owe honour for <u>what</u> you have done, and for the fact that you did it for <u>him</u>.

From the depths of my heart I wish you consolation and peace, and greet you sincerely,

A. Einstein. [111]

15

Marcel

The spacious campus of Rome's *Università La Sapienza* lies in the glittering light of a midsummer day; heat shimmers above the lanes. The cicadas are singing loudly in the high pine trees, and they drown out even the footsteps of the many people who are walking from building to building, leisurely dressed, chatting with each other, or lying on the lawn in the shadows of the trees. They fan themselves with one hand and hold a plastic fork in the other, eating from their lunch boxes. They are all wearing cherry-red ribbons with name tags; they are speaking English, Italian, Portuguese, German, French, Spanish, Polish, Czech, Chinese, Hindi, Hebrew, Norwegian, Romanian, Swedish, Vietnamese or Farsi.

The date is July, 2015; nearly eighty years have passed since Marcel Grossmann closed his eyes forever in the Paracelsus Clinic in Zurich. And yet his name is omnipresent during this week in Rome. It is emblazoned on varicoloured flags, on glowing red T-shirts, on backpacks and countless name tags. Young and young-at-heart scientists from fifty countries have travelled to the Tiber to participate in the 14th Marcel Grossmann Meeting. The thirteen previous conferences took place with a three-year rhythm in Trieste, Shanghai, Rome, Perth, Kyoto, Stanford, Jerusalem, Rio de Janeiro, Berlin, Paris, and Stockholm. What moved Professor Remo Ruffini, the charismatic astrophysicist and director of ICRA and ICRANet, to name an international meeting series after Marcel Grossmann forty years ago? Ruffini formulates it as follows: "Abdus Salam and I wanted to emphasise the importance of the man who built a bridge between physics and mathematics that still inspires us today." The scientific goals of the conference series

© Springer International Publishing AG, part of Springer Nature 2018
C. Graf-Grossmann, *Marcel Grossmann*, Springer Biographies,
https://doi.org/10.1007/978-3-319-90077-3_15

are thus to bring together scientists from a variety of different special fields in order to deepen our understanding of the structure of spacetime and to take stock of the various experiments which relate to Einstein's theory of relativity. The spectrum reaches from abstract classical theories of gravitation to quantum physics and string theory to relativistic astrophysics and to operating and planning observational missions and instruments.

Evidently, the meetings have aroused considerable interest: From July 12th to 18th, 2015, over 1000 lectures and classes are held in Rome, and around 1200 ladies and gentlemen—mathematicians, physicists, astrophysicists, historians of science, specialists in relativity theory and chemists, among them several Nobel Prize winners—have come to celebrate the centennial of General Relativity and the golden jubilee of relativistic astrophysics, as well as the UNESCO Year of Light. In the auditoriums and classrooms of the various university institutes, they present their topics, projects and formulas. The Theory of General Relativity, which will be 100 years old in November, is most especially at the focus of this meeting.

What motivates the participants to travel long distances to attend these meetings? Prof. Philipp Jetzer from the Physics Institute of the University of Zürich sees it thus: "I have been attending the MG Meetings regularly since 1988 (i.e. since the fifth Meeting in Perth). My reasons are on the one hand to keep up with research by participating in the lectures; on the other, to maintain personal contacts with other participants, which are very important. We can exchange and discuss new ideas. I have also met many people in this way with whom I collaborated in the following years, or am still doing so. Generally, taking part in regular conferences is very important for our work." Helena Durnová, Ph.D., professor at the Department of Mathematics of the Masaryk University in Brno, Moravia [Czech Republic], is working at present on a biography of the Czech mathematician Václav Hlavatý. He worked in the 1950's and 60's with the mathematical apparatus which was intended to support a unified field theory. "For me, the Marcel Grossmann conference which brings together a multiplicity of scientists who are working on the physical and mathematical aspects of relativity theory (including those who are working in the field of the history of physics) is a very good opportunity to meet colleagues and discover possible new connections. I hope that this will help me to place the career of Václav Hlavatý in a wider context."

At every Meeting, prizes are awarded, one to an institution and others to individual scientists. This year, the European Space Agency, ESA, is to be awarded the institutional prize, whilst Professors Ken'ichi Nomoto, Lord

Martin Rees, Yakov G. Sinai, and Sachiko Tsuruta are the winners of the prestigious personal awards. All of them are given a sculpture created by the Italian artist Attilio Pierelli, whose original was presented to Pope John Paul II in 1985, in recognition of the achievements of the Vatican Observatory at Castel Gandolfo.

Change of scene. Wednesday evening; tonight, the festive banquet at the midpoint of the conference is taking place. Above the spacious inner court-yard of the *Palazzo Colonna* in the inner city of Rome lies a soft, peach-coloured evening light. The humid heat which lay heavily upon the Eternal City all day long is gradually lessening and making way for cooler night air. At the festively set tables in the Baroque garden, with its lemon trees and accurately trimmed boxwood shrubs, scientists from all over the world are sitting with their partners and colleagues. Light-coloured shirts shine, delicate summer dresses rustle, people are chatting and laughing. Following the first three intensive days of the Meeting, some relaxation in an elegant setting is more than welcome.

A few hours earlier, 137 guests of honour, including diplomatic, cultural and religious dignitaries, had gathered in a splendid town house, the Renaissance *Palazzo Besso*, which had once belonged to the Florentine *Strozzi* family and is today the seat of the renowned Marco Besso Foundation. In the ancient and dignified rooms, amid valuable books and works of art, the guests commemorated the collaborative efforts of Albert Einstein, Michele Besso, and Marcel Grossmann. Professor Hanoch Gutfreund, President of the Hebrew University of Jerusalem for many years, and Professor Jürgen Renn, Director of the Max Planck Institute for the History of Science in Berlin, presented their recent works which were published by the Princeton University Press, "*The Road to Relativity*", and "*Relativity, The Special & The General Theory, 100th Anniversary Edition*" to the interested, informed and sympathetic audience. They have also curated an exclusive exhibition on the collaboration between Einstein, Besso, and Grossmann. Then the author of this biography had the opportunity to depict the life of her grandfather, and thus to introduce the person whose name was mentioned so often during this week.

In the meantime, the warm Roman night has fallen on the *Palazzo Colonna* and its inner courtyard. The elegant festive evening is slowly winding down; on the following day, all the attention of the participants will once again be focussed upon scientific discussions and lectures. Until Saturday, Marcel Grossmann's name and his spirit will still continue to be very much alive and present.

What conclusions can we draw today, eight decades after his death, about the life and work of the mathematician? On the pages that follow, Prof. Tilman Sauer, historian of science, physicist and member for some years of the Einstein Papers Project, subjects Marcel Grossmann's academic achievements to a critical appraisal.

Epilogue—A Scientific-Historical Assessment

Marcel Grossmann was a mathematician through and through. His interest in mathematics must have already arisen during his school days. Although his School Leaving Certificate from the Higher Middle School [*Obere Realschule, Gymnasium*] in Basel in fact attests to excellent marks in all of his subjects [112], it mentions explicitly, in the column for 'Disposition', that Grossmann plans to attend the Polytechnic Institute and to study mathematics there [113].

Department VIA of the Polytechnic Institute (now the ETH Zurich), i.e. the "Mathematical Section" of the "School for Specialised Teachers in Mathematics and Scientific Areas", required a challenging obligatory basic curriculum in mathematics of its students in their first two years. This started off with a course in differential and integral calculus and differential equations, given by the chairman of the Section, Adolf Hurwitz (1859–1919). Then there was a lecture course in mechanics given by Albin Herzog (1852–1909), whilst Carl Friedrich Geiser (1843–1934) gave the lectures on analytic geometry and determinants. Of particular importance for Grossmann were the lectures of Otto Wilhelm Fiedler (1832–1912).

From the latter, he attended a course in descriptive geometry in his first year, along with another on projective geometry. A non-compulsory course on central projections, also given by Fiedler, is notable; Grossmann attended it in his first semester, as well as a lecture series on "winding" curves which he attended in his fourth semester.

© Springer International Publishing AG, part of Springer Nature 2018
C. Graf-Grossmann, *Marcel Grossmann*, Springer Biographies,
https://doi.org/10.1007/978-3-319-90077-3

PROF. DR. OTTO WILHELM FIEDLER
1832–1912

Photos The *Higher Middle School [Gymnasium]* in Basel, on a postcard from 1905 | Otto Wilhelm Fiedler, Grossmann's professional mentor and teacher at the Polytechnic Institute

It appears that along with his interest in mathematics itself, Grossmann also developed an interest in the teaching of mathematics. Among the papers in his estate, one finds a report on the instruction in mathematics which he received at the Higher Middle School in Basel [114]. This report is addressed to Professor Fiedler and dated July, 1898; thus at the end of his fourth semester. In this report on the mathematics education in Basel, Grossmann describes in some detail the content and the methods of teaching. He added some reflections on the teaching of mathematics to his description of the material taught. Here, he took up a topic of interest to Fiedler, starting off by deploring the lack of unity and uniformity in the presentation of the material:

> In fact, the pupils do not receive a unified impression of geometry; to the gifted student, it appears to be a collection of all the possible, randomly-discovered theorems, whilst those who are less able to keep up are doomed to drown in the chaos (whole volumes are dictated).

The young Grossmann thus began already during his student years to demand a reform in the teaching of mathematics:

> However, I believe that it would have to begin very early; in our middle schools, we have for the most part the poorest class of teachers of mathematics, under whose instruction even the most beautiful subject would wither and dry up, & under whom one of the most elegant aspects of mathematics, its exactness in all directions, is distorted into pedantry. 90% of the pupils acquire a feeling of revulsion towards this dry mathematics, & their interest in it can hardly, or only seldom, be aroused in the higher forms, even by the best system. That one can awaken pleasure and interest in mathematical science even in average students, not to mention the more gifted ones, has been my own experience with pupils to whom I have given private tutoring or preparatory lessons for an entrance examination. One must simply know how to motivate them.

The lectures in his first two years of study were for the most part obligatory for all of the students in Department VIA, and a specialisation to a particular mathematical or physical direction began only in the third or fourth year. In these last two years at the Zurich Polytechnic, Grossmann specialised in mathematics, taking lecture courses on elliptical functions, algebra, variational calculus and partial differential equations from Hermann Minkowski (1864–1909), on complex numbers from Adolf Hurwitz, on planar curves and algebraic surfaces from Carl Friedrich Geiser, and on linear differential equations and the theory of invariants from Arthur Hirsch (1866–1944). His lecture notes from these courses, worked over with elaborate care, can still be seen today by anyone who is interested, in the scientific-historical collections of the ETH, either as the originals or, more recently, online via the e-manuscript portal of his former college. As a whole, they form a fascinating, rather complete picture of mathematics instruction at one of the best technical colleges around the turn of the past century.

More importantly, however, Grossmann participated regularly in the mathematical seminar led by Geiser, Minkowski and Fiedler, and was introduced to current research topics by those leading representatives of their respective fields. He felt especially attracted to the research areas of his teacher Wilhelm Fiedler. Fiedler's area of specialisation was geometry, in particular descriptive geometry and projective geometry.

In Grossmann's time, geometry had a different significance in the canon of the mathematical disciplines and in teaching than it does today, especially for the technical education at the Polytechnic. Technical professions such as mechanical engineering, machine design and construction, and architecture require accurate, precise and reliable descriptions of their objects in technical drawings, plans and sketches. These representations show the three-dimensional objects from different viewpoints as top [plan], front and side views, and they are thus brought into perspective. What today is accomplished by essential digital tools such as computer-aided design (CAD) and other programmes still depended on hand-drawn plans and technical drawings in the not-too-distant past, and certainly in Grossmann's time. The teaching of descriptive geometry thus had a particular importance at the Polytechnic.

Geometry served however not only for the technical instruction of practitioners in handling such methods of representation. It was in addition a basic mathematical area in its own right. Its origins go back to Euclid and his famous *Elements*, which had been taught in the schools and universities of Europe for centuries, and which formed our concept of geometry. The constructions of circles, triangles, squares and other polygons in the plane, as well as the three-dimensional figures of spheres, cones, cubes etc., and on to the Platonic Solids, as they were described in the *Elements*, are still a regular component of school mathematics today. By the beginning of the 20th century, however, the ancient and honourable discipline of geometry had already evolved considerably. Geometry at that time was divided into various sub-disciplines. In order to be able to appreciate the work of Marcel Grossmann, we shall find it necessary to briefly consider these various aspects of the science of geometry.

Out of descriptive geometry, projective geometry had developed. Its goal is to investigate the theoretical relationships of perspective representations in terms of their internal relationships. In perspective representations, for example those generated by a central projection, distances appear to be foreshortened according to the perspective, and angles are distorted. At the same time, infinitely distant points are represented, such as those familiar as the vanishing points in artists' perspective paintings and drawings.

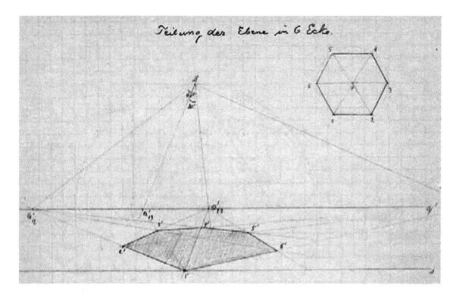

Figure Construction of the perspective representation of a hexagon; drawing by Marcel Grossmann in his lecture notes from a course on central projections given by Otto Wilhelm Fiedler in the winter semester of 1896/97

In the second half of the 19th century, as a culmination of projective geometry, the "geometry of position" had become established. In this subfield, the goal was to investigate only those geometric relationships in which no metric concepts may be used, i.e. no distances and no angles, but instead only the relative positions of points, lines and planes.

Along with these subfields of synthetic geometry, in which constructions are made using a ruler and a compass, there were also developments in other directions. One of these was related to the axiomatic foundations of geometry. Euclid's *Elements* was the model example, valid over thousands of years, of a deductive science in which, starting from only a few postulates, axioms and definitions, every other theorem can be deduced from the previous theorems and insights by means of logical sequences, and can be proved. Among Euclid's fundamental axioms, he had included one which was called into question again and again by geometers over the centuries, the so-called parallel axiom. It states (in one of its various possible formulations) that within a plane, one and only one straight line can be drawn through a point which does not lie on a given line in such a way that the two lines never intersect. It had been uncertain for mathematicians since Euclid's time whether this axiom is indeed independent of all the other axioms, or perhaps can be deduced from them. In the 19th century, through the works of Carl Friedrich Gauss (1777–1855), János Bolyai (1802–1860), Nikolai Lobachevski (1792–1856), and Bernhard Riemann (1826–1866), it finally became clear that the parallel axiom is in fact independent. It could indeed be shown that there are consistent geometric systems in which this axiom can be replaced by different statements. For example, there are geometries in which, through a given point outside a given line, either many other lines can be drawn which do not intersect the given line (hyperbolic geometry), or else none at all (elliptical geometry). These so-called non-Euclidean geometries therefore themselves became the objects of geometrical research.

From the investigation of curved surfaces in space, Gauss had founded another direction within geometry, in which such curvature relations are studied independently of their embedding in three-dimensional space (Gaussian theory of surfaces). The continuing question as to what extent such curvature relations can be extended to three and more dimensions (differential geometry) was also studied. For the treatment of these latter questions, one naturally had to employ coordinate representations and coordinate transformations between them, such as had been developed in analytic geometry since the time of René Descartes (1596–1650). Here, multi-dimensional differential geometry could profit from the more analytic parts of mathematics, such as the theory of invariants.

This multiplicity of geometric subfields of analytic, descriptive and projective geometry as well as synthetic geometry was reflected in the curriculum of the Polytechnic Institute in Grossmann's time, and it forms the scientific-historical background not only of his mathematics education and his own further research, but also of his later collaboration with Einstein.

After obtaining the Swiss Diploma for Specialised Subject Teachers in Mathematics in July of 1900, Grossmann held an assistantship with his professor and mentor Fiedler, in order to write his doctoral thesis under Fiedler's supervision. Correspondence with Fiedler in the ETH archives shows that Grossmann was advised by Fiedler about the choice of a topic for his dissertation. Already after only two years, Grossmann submitted his thesis, in which he dealt with a special topic from projective geometry.

Grossmann's further research work dealt with the question of how one can make non-Euclidean geometries clear in the sense of descriptive geometry. This problem thus touched upon the internal connections of two relatively independent sub-disciplines within geometry. Arthur Cayley (1821–1895) and Felix Klein (1849–1925) had described a method in the second half of the 19th century with which the different metric relationships in non-Euclidean geometries could be mapped projectively, and thus the usual Euclidean construction exercises on paper could be made possible for the non-Euclidean geometries as well. The further research works of Grossmann in the following years dealt with the solution of this problem. In particular, Grossmann completely solved the problem of how to treat the classical Euclidean construction exercises for the case of hyperbolic and elliptical geometries in the framework of the Cayley-Klein formalism. He published these construction exercises in complete form in a supplement to the programme of the Thurgau Canton School for the school year 1903/04 [115]. This form of publication, unusual from today's standpoint, permitted the teachers at the Canton School to publish their research results in a relatively detailed form and in a mannerly edition. Grossmann also published one of the most difficult and challenging of his construction exercises in the respected professional journal *"Mathematische Annalen"* [Annals of Mathematics], in addition [116].

An impression of the complexity of this construction is given by the figure on the following page, in which the exercise of constructing a straight-sided right triangle whose angles are known is solved. The interior of the circle denoted by Ω (conic section) represents hyperbolic geometry here, whilst the circle itself represents all infinitely distant points. The triangle that is to be constructed has a right angle at C, which in this representation of hyperbolic geometry however does not appear as a Euclidean right angle. Furthermore, the angles α and β are given, and the solution of the exercise consists in constructing the points A and B.

His research on projective constructions in non-Euclidean geometries exhibits a certain complexity and demonstrates that Grossmann had knowledge of the most advanced geometrical theories, even if this research was hardly relevant to the education of students in the schools or even at the Polytechnic Institute at the time. Grossmann's work dealt with fundamental research in geometry. In an historical overview of the history of non-Euclidean geometry, written by the Italian mathematician Robert Bonola (1874–1911), in its German translation from 1908 by Heinrich Liebmann (1874–1939), Grossmann's work is mentioned right at the end as representing the current state of research [117].

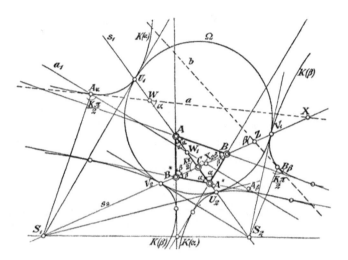

Figure The construction of a right triangle in an hyperbolic geometry

Liebmann himself worked in this field, and mentioned Grossmann's approach, which went beyond his own work, and especially the *"tools of projective geometry, with the aid of which M. Grossmann has carried out his beautiful constructions"* [118], in an article on "Elementary Constructions in the non-Euclidean Geometries" from the year 1911. In a review article on "Elementary Geometry and Elementary non-Euclidean Geometry in a Synthetic Treatment" in the comprehensive *"Encyclopädie der mathematischen Wissenschaften"* ["Encyclopedia of the Mathematical Sciences"] from 1913, Max Zacharias (1873–1962) mentions Grossmann's work [119]. In his description, however, Grossmann's method of construction appears as an unorthodox alternative to more direct procedures in which the projective methods are not used [120].

Based on these research results on non-Euclidean geometry, Grossmann obtained his *Habilitation* in 1905 at the University of Basel. At that time, he was teaching at his earlier high school, the Higher Middle School [*Gymnasium*] in Basel. As an aid to his instruction, he wrote two textbooks in the year 1906, one on analytic geometry [*Analytische Geometrie—Leitfaden*] and one on descriptive geometry [*Darstellende Geometrie—Leitfaden*] [121]. His earlier thoughts as a student on the quality of the mathematics instruction in Basel may have played a role in his writing of these guiding textbooks ['*Leitfäden*'], but a new curriculum plan approved for the schools in Basel in 1903 was also a motivating factor. In the foreword to his little book on analytic geometry, Grossmann writes:

> The writing of the present guide was motivated by a desire to provide the students at the Higher Middle School in Basel with a summary of the material treated in their classes, so that their teacher is spared from lecturing, and the pupils are spared from taking notes. In this way, time is gained which can be put to good use in intensive practice and application of the so fruitful methods of analytic geometry.

An historical background for Grossmann's geometry textbooks was the discussion on the requirements for the teaching of geometry which was raging at that time at many schools and universities. Fiedler's major textbook on "*Darstellende Geometrie in organischer Verbindung mit der Geometrie der Lage*" ["Descriptive Geometry in a Systematic Combination with Positional Geometry"] had appeared as the third edition in three volumes in the years 1883–1888 [122]. With more than a thousand pages, it represented a standard work in its field. Fiedler's adaptation of Salmon's textbook on the "*Analytische Geometrie des Raumes*" ["Analytic Geometry of Space"], the so-called "Salmon-Fiedler", had also appeared in 1898 as two thick volumes [123]. But Fiedler had shown himself in his lecture courses at the Polytechnic to be disinclined to take the practical needs of engineering students into consideration, which occasionally led to conflicts and altercations. All the more were his extensive textbooks unsuitable for use for the instruction in schools, and there was a need for more brief treatments which contained mainly the material relevant to school mathematics or to the practical education of engineers [124].

Einführung

in die

Darstellende Geometrie.

-i - i-

Leitfaden

für den Unterricht an höheren Lehranstalten

von

Dr. Marcel Grossmann

Prof. an der Eidg. Technischen Hochschule in Zürich.

Dritte Auflage, mit 85 Übungsaufgaben
und 121 Figuren in besonderem Heft.

Basel 1917
Verlag von HELBING & LICHTENHAHN.

Photo Title page of Grossmann's textbook on descriptive geometry

Grossmann's geometry textbooks and his own research qualified him for a position at the Polytechnic Institute itself. When Fiedler had to be substituted for a semester due to his advanced age, it was only natural that Grossmann took over his duties. A short time later, Fiedler had to retire from his professorship altogether for health reasons, and Grossmann became his successor. When Fiedler died in 1912, Grossmann wrote an epitaph for him [125].

After Grossmann had obtained the position as successor to Fiedler at the Polytechnic Institute, he increasingly took on more administrative and scientific-organisational duties. A comprehensive report on mathematics education at the Polytechnic, which in 1911 was renamed the '*Eidgenössische Technische Hochschule*' (ETH), or Swiss Federal Technical Institute, is especially worthy of mention. It was written by Grossmann as part of the report on the educational situation in Switzerland for the International Commission on Mathematics Instruction [126].

As the chairman of the (newly renumbered) Section VIII for Specialised Teachers of Mathematics and Physics, beginning in the winter semester 1911/12, Grossmann was also actively involved in bringing Einstein to the ETH as a colleague in his department. When Einstein arrived in Zurich in August of 1912, a close collaboration began between the two friends, which led to Grossmann's most widely known publications. Within this collaboration, Grossmann raised Einstein's efforts to create a generalised theory of relativity onto a new level, by introducing him to the mathematical field of tensor calculus. With a certain justification, Grossmann can therefore be called the father of the mathematical formulation of General Relativity; this impact of his work is recognised today by the fact that one of the major international conference series on current research in General Relativity theory, gravitation, and astrophysics [the Marcel Grossmann Meetings] bears his name.

What exactly was Grossmann's contribution? [127] To answer that question, we have to take a closer look at just what Einstein's research problem was when he turned to his old friend in the summer of 1912 with a request for help on a problem with which he could go no further. In his *annus mirabilis* in 1905, Einstein had formulated the theory of Special Relativity, in which the fundamental concepts of space, time and velocity had already undergone deep-seated revisions. In this theory, there is no longer an observer-independent concept of simultaneity. The combination of two velocities is also no longer obtained through simple [vector] addition, as in classical mechanics.

But Special Relativity refers strictly speaking only to the relativity between frames of reference which are moving relative to one another with linear and constant velocities [so-called *inertial frames*]. Furthermore, Special Relativity contradicts the time-honoured Newtonian theory of gravitation, in that it requires that no physical effects can propagate faster than the finite speed of light, whilst Newton's gravitational attraction, a force-at-a-distance, propagates infinitely fast. When Einstein was confronted with the necessity of making a statement about this problem when he was writing a review article on his new theory in 1907, he formulated a hypothesis with which he hoped to resolve the contradiction. Since the time of Galilei, it had been known that all bodies fall at the same speed. Einstein now presumed that Galilei's principle of a constant speed of falling holds universally, and he founded his hypothesis of *complete* equivalence between acceleration and gravitation upon it. He assumed that in principle, it is never possible to distinguish between a space without a gravitational field which is however being uniformly accelerated, and a space which is at rest but where a homogeneous and static gravitational field is acting. Speaking pictorially: If one is inside a closed room which is not too large, then it is impossible to decide whether one is for example on the surface of the Earth in the constant gravitational field of the planet, or is floating in open space, but being accelerated upward at a constant rate by a rocket drive. In both cases, one will namely be "pressed down toward the floor", and in both cases, all objects fall at the same rate.

Making use of this equivalence hypothesis, Einstein was able to theoretically investigate the gravitational field by considering field-free but accelerated systems. As long as he limited his considerations to the case of uniform, linear acceleration or uniform rotation, he could make good progress with his assumptions, and could even predict that in a gravitational field, light beams would be curved, and that the light which is emitted by atoms on the Sun and propagates to the Earth should have a somewhat different colour [i.e. a shift of its spectral lines] from light emitted by similar atoms on the Earth's surface. The point beyond which Einstein could not proceed was when he considered how the theory should look in the case of arbitrary accelerations; for example, when he tried to calculate the forces which act on the passenger in a wildly and irregularly moving bumper car.

Mathematically, the description of such generalised states of motion leads to the problem of general transformations on arbitrarily curved coordinate systems. This problem had already been treated in detail by mathematicians in the framework of the theory of invariants. Grossmann, as a specialist on non-Euclidean and descriptive geometry, was himself not an expert in the field of the theory of invariants, but he had nevertheless had such a good mathematics education that he rapidly deduced where the key to Einstein's problem must lie. He read the corresponding literature, among other things an important paper from 1869 by the founder of the Mathematical-Physical Section of the Polytechnic, and thus his predecessor as chairman of that Section, Bruno Elwin Christoffel (1829–1900), which dealt with this question [128]. Christoffel's work already contained the most important elements for the solution of the problem, but it was written in a manner that was difficult to access. Grossmann therefore continued his search and found a paper by Bernhard Riemann, which likewise already contained the decisive mathematical pieces of the puzzle, but was also not very accessible [129]. Continuing his literature search, Grossmann found a textbook on differential geometry by Luigi Bianchi (1856–1928) in the German translation by Max Lukat [130], which contained a summary of Christoffel's results; and finally, he found an article by the Italian mathematicians Gregorio Ricci-Curbastro (1853–1925) and Tullio Levi-Civita (1873–1941), in which they had already worked up and formulated the required mathematics as a calculus [131]. Ricci and Levi-Civita called their formalism "*absolute differential calculus*", and Grossmann recognised that this calculus already provided almost precisely the techniques which were needed to solve Einstein's problem. However, the application of the "absolute differential calculus" still required an additional adaptation by Grossmann and an interpretation as a tensor calculus, before Grossmann and Einstein succeeded in formulating the problem of a generalised theory of relativity with the aid of this mathematical language, which had been completely unknown to Einstein and all the other physicists of that time.

Since Grossmann and Einstein could see each other daily at the time, their scientific collaboration was close and intense; at the same time, we unfortunately know very little about what the two researchers discussed and investigated. There is no correspondence at all between the two friends which has survived from this time, and also no notes by Grossmann. There is, however, an important document from Einstein's estate, which gives information about their search for a relativistic theory of gravitation, and in particular their search for the correct gravitational equations. This is Einstein's so-called *Zurich notebook*, a bound notebook containing around 80 pages, with computations and notes in Einstein's handwriting from the time between summer 1912 and summer 1913 [132]. Although it was written by Einstein alone, it still contains the relevant indications about Grossmann's role in their collaborative work. At two decisive points, Einstein mentions the name of his collaborator.

On page 14L in the Zurich notebook, one finds for the first time a note on the Riemannian curvature tensor, and Einstein writes next to it, "Grossmann tensor of fourth order" (see the figure at upper right).

Einstein's note looks innocuous, but it has historical significance. At this point, namely, the physical problem of the formulation of a relativistic theory of gravitation, and the concepts of a mathematical tradition with a corresponding well-developed calculus, were brought together for the first time. With this decisive step, all the prerequisites were indeed in place so that Einstein and Grossmann could have formulated the correct field equations and thus could have competed the theory of General Relativity.

What can also be seen in the excerpt from the notebook shown is that Einstein had started here with a mathematical re-formulation which leads from the Riemannian tensor to the so-called Ricci tensor, which Ricci had in fact already investigated in 1904, but which had not yet been introduced into the literature that Grossmann had presumably studied. A more careful reconstruction of the calculations in the notebook in fact demonstrates that Einstein and Grossmann, a few pages later, had already written down the correct equations in the linear approximation. Nevertheless, they soon abandoned those equations and pursued other possibilities. One of these is connected with Grossmann's name by Einstein (see the figure below).

This second excerpt also appears to be innocuous at first glance, but it is indeed of great interest. The quantity which is in fact constructed here, and which is mentioned by Einstein as "presumably the gravitational tensor", is precisely the quantity that he considered again three years later for the gravitational equations and which then permitted his final breakthrough to General Relativity theory.

Figures <u>Above</u> | On this page of Einstein's Zurich notebook, the Riemannian curvature tensor is mentioned for the first time; it is the central starting point for the mathematical formulation of the theory of General Relativity. Einstein notes it here with the remark, "Grossmann tensor of fourth order"
<u>Below</u> | On this page of the Zurich notebook, Grossmann shows the possibility of splitting a quantity which is termed here "the presumed gravitational tensor" off the Ricci tensor. Precisely this quantity was the starting point from which Einstein, three years later and without Grossmann, began his return to generalised covariance

But in the Zurich notebook, this possibility is also rejected. Instead, we find a few pages later the sketch of a derivation of the field equations which no longer makes use of the Riemannian tensor and its possibilities at all.

Grossmann's introduction of tensor calculus could thus have already brought the search for a generalised theory of relativity and a relativistic gravitational theory to a successful conclusion. All of the elements of the later theory of General Relativity, and thus of our current understanding of space, time and matter, and of the structure of the Universe in general, were in principle present. However, Einstein and Grossmann did not succeed at that time in achieving the final formulation of the theory. The last, decisive step was still missing. The new mathematical formulation was based on the introduction of the metric tensor, a complex of ten functions with particular transformation properties, which on the one hand describe the generalised geometry of spacetime, and on the other are supposed to represent the gravitational potentials. In order to bring this theory to a consistent conclusion, it therefore still required equations with which the components of the metric tensor could be determined as functions of the distributions of matter and energy in space. These equations should lead again to the well-known and tested equations of Newton's theory in the limiting case of Newtonian gravitation, and this was the source of the difficulty. Einstein and Grossmann did not succeed in recognising how this limiting case was to be arrived at, even though they were already looking at the correct equations.

After several months of intense collaboration, the two friends had indeed given up their original goal of finding a completely covariant theory, but they had on the other hand found equations which seemed to essentially fulfil all of the requirements. They decided to write up the results of their collaboration and leave them open to public discussion. In early 1913, they wrote an "Outline of a Generalised Theory of Relativity and of Gravitation". This presentation consists of two parts, a "physical part" for which Einstein was responsible, and a "mathematical part" written by Grossmann. As the name already implies, this publication describes an unfinished theory, but Grossmann's mathematical part had introduced precisely the mathematical elements of a tensor calculus which proved later to be the decisive instruments for the formulation of the final theory [133].

A more precise examination of Grossmann's "mathematical part" shows that this is itself divided into two parts. In the first three paragraphs, Grossmann draws up the fundamentals of a calculus which one can readily recognise from today's viewpoint as tensor calculus, and which forms the basis of all the calculations in the theory of General Relativity. Although Grossmann refers explicitly in the beginning to Christoffel's work and to the absolute differential calculus of Ricci and Levi-Civita, his description is thoroughly original and modifies the earlier calculus in several important points. Thus, for example, Grossmann introduces the term "tensor" here for the first time for the quantities which were still called "systèmes" by Ricci and Levi-Civita, a not unimportant step, for these mathematical quantities were thus connected with a physical interpretation. Furthermore, Grossmann makes some changes in the notation. A remark of Grossmann's at the end of his introduction is notable:

> Here, I have intentionally left out geometrical methods, since they in my opinion would contribute little to the clarification of the concepts of vector analysis. [134]

It is seen in fact that the mathematics used by Einstein in the formulation of the theory of General Relativity is mainly of analytical nature, and Einstein never explicitly realised the geometric significance of the formulation of the theory. Important geometric interpretations were found only after the completion of General Relativity, by mathematicians such as Hermann Weyl (1885–1955) and Tullio Levi-Civita, with the concept of an affine connection. That Grossmann intentionally dispensed with geometrical methods in his collaboration with Einstein is interesting, considering his professional expertise in descriptive geometry, as it indeed shows his precise knowledge of the state of the literature—since a geometrical illustration, such as would have been feasible at this point, would have probably not have made possible any really new insights.

Following the publication of the "Outline", the two friends continued with their formulation of the theory and presented it at various scientific gatherings. A year later, they achieved a further important milestone: They were able to show how the field equations of their "Outline" could be derived by another route, namely using the methods of variational calculus. That to be sure did not make those equations correct, but one could recognise some of their properties more clearly, in particular those resulting from their restricted transformation group. Einstein and Grossmann published their new derivation in a second article, this time really written jointly, with the title "Covariance Properties of the Field Equations of the Theory of Gravitation based on the Generalised Theory of Relativity" [135]. This second joint article appeared in May, 1914, shortly after Einstein had already left Zurich in order to take up his new position as a member of the Prussian Academy of Sciences in Berlin.

The distance between Zurich and Berlin brought an end to the collaboration between the two friends and former colleagues. When shortly thereafter the War broke out, a cooperation across the borders became even more difficult, in particular because Grossmann now had to devote himself increasingly to his duties as a teacher at the ETH. From then on, Einstein worked alone on his theory, and 18 months later, he achieved a breakthrough and the greatest success of his scientific career. In a dramatic twist in the fall of 1915, he saw where the errors had been made in the earlier theory of the "Outline", and in a series of four papers published at intervals of one week in November 1915, he returned to the general covariance theory; in the last of these papers, he formulated the final version of the field equations of the theory of General Relativity. In the introduction to the first paper in this famous November series, Einstein summarises his insights into the errors in the "Outline" theory and continues:

> For these reasons, I lost trust in the field equations I had formulated, and instead looked for a way to limit the possibilities in a natural manner. In this pursuit I arrived at the demand of general covariance of the field equations, a demand from which I parted, though with a heavy heart, three years ago when I worked together with my friend Grossmann. As a matter of fact, we were then quite close to that solution of the problem which will be given in the following. [136]

It proved that the theory of the "Outline" which had been formulated by Einstein and Grossmann needed only to be formally re-interpreted at a few places in order to obtain, nearly unchanged, the final theory. In this process, the variational method introduced in the second joint paper played an important role.

After the end of the collaboration with Einstein, Grossmann returned to his duties as professor and mathematician. These duties now demanded all his energies, especially since the situation of education at the universities in Switzerland had changed drastically after the outbreak of war. Another task also began to occupy a considerable part of his energy even during the War, namely the efforts to reform the standards of the Swiss School Leaving certificates. Grossmann was an academic teacher and supporter of instruction in mathematics with heart and soul, and he educated generations of students at the ETH in analytic and descriptive geometry. For that purpose, he made use of the textbooks that he had written, which appeared in several new editions during his tenure as a university teacher, and which he continually revised and extended. An overview of Grossmann's textbook publications is given in the following outline.

His "guides" ["*Leitfäden*"] to analytic geometry (AG) and to descriptive geometry (DG) from 1906 were written especially for use at the Higher Middle School in Basel, i.e. they were aimed at middle schools. The guide to descriptive geometry was reissued as a second edition in 1912, and as a third in 1917. By that time, however, Grossmann already needed a textbook for his lecture courses which would be more suited for instruction at the ETH, and thus at technical colleges. Therefore, in 1915 he wrote a textbook on descriptive geometry which offered to *"the students"* at technical colleges and universities *"a compact accompaniment to aid their studies"*. This new book was published by Teubner in Leipzig and Berlin, and for reasons of compactness, it intentionally omitted the *"elements"* of descriptive geometry which *"belong in preparatory instruction in schools"*. It however soon became clear that there was a demand for a separate account of these "elements", and two years later (1917), Grossmann published a small volume on the "Elements of Descriptive Geometry", likewise issued by Teubner, as a complement to his textbook for the university level from 1915. When after the end of the War a second edition was called for, Grossmann combined the "Elements" as Part I, and the previous textbook as Part II, into an Introduction to Descriptive Geometry. This second part was again modified by Grossmann in 1932 into a self-contained introduction to descriptive geometry with the subtitle "Mapping Procedures, Curves and Surfaces". Finally, it should be mentioned that in the year 1927—after twenty years of teaching in the field of descriptive geometry for engineers—Grossmann again wrote a "Descriptive Geometry for Mechanical Engineers", published by Julius Springer in Berlin.

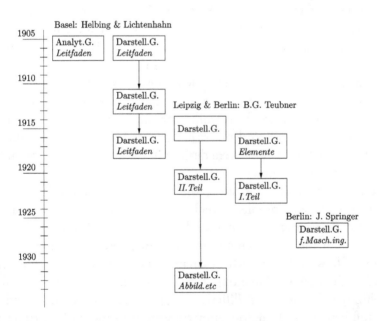

Figure Chronological outline of Grossmann's textbook publications as a schematic overview. <u>Abbreviations</u>: Analyt. G. *Leitfaden* = Guide to Analytic Geometry; Darstell. G. *Leitfaden* = Guide to Descriptive Geometry; Darstell. G. *Elemente* = Elements of Descriptive Geometry; *I. Teil* = Part I; *II. Teil* = Part II; *f.Masch.ing.* = for Mechanical Engineers; *Abbild.etc* = Mapping Procedures, etc.

Grossmann's various textbooks mirror the different and sometimes contradictory requirements of practically-relevant instruction in geometry, which was the subject of continuing discussions. In the foreword to his "Descriptive Geometry for Mechanical Engineers" from 1927, he hints at what is behind these textbooks:

> Those who teach descriptive geometry will know that external conditions often make it difficult to realise didactic principles within their instruction. Especially at smaller and medium-sized technical colleges, the students participating in such courses tend to be an extremely heterogeneous group, consisting of future architects, civil and surveying engineers, mathematicians and physicists, whose needs vary considerably within the group. The present volume thus does not reflect any particular university-level lecture course, but rather offers an overall image of such courses in general, such as I have given since 1907 for the mechanical engineers at the Swiss Federal Technical Institute in Zurich. [137]

With Grossmann's commitment to mathematics education, his journalistic engagement for the "*Neue Schweizer Zeitung*" ["New Swiss Journal" newspaper] [138], and his efforts on behalf of the reform of Swiss School Leaving certification, it is not surprising that his publications in mathematics research begin again only after the end of the War; however, by this time, his health situation was already causing him worries and difficulties.

His works from his later years are again characterised by an instructive-geometrical approach. And these works contain a clearly stronger relevance for engineering and technical practice. Almost all of his research also treats questions whose foundations were dealt with in his lecture courses and textbooks on descriptive geometry.

As early as 1910, in a lecture to the Swiss Mathematical Society, Grossmann had discussed a problem which is to some extent the inversion of the problem of perspective representation [139]. There, in fact, he dealt with the question of solving the geometric problem of photogrammetry, that is the question of how and under which circumstances and with which geometric methods a three-dimensional object can be reconstructed from several photographic registrations from different perspectives. The elements and theoretical fundamentals of photogrammetry were the subject of a chapter in his textbooks from 1915 on.

In 1925, he published an article on the "complete focal system of planar algebraic curves" [140]. The basic geometrical concepts of the treatment of curves, such as the definitions of their orders and classes, their tangents, osculating planes and focal points, were likewise treated within descriptive geometry, although for example the theory of "winding curves" had been the subject of its own lecture course during his student days, and the treatment of curves and surfaces of higher order and class is a very extensive field of geometrical research in itself.

A year later, Grossmann was similarly occupied with the problem of a geometric curve; in fact a special three-dimensional curve of third order, the so-called horopter curve [141]. This curve plays a fundamental role in the theory of three-dimensional binocular vision. There, one indeed finds that when the gaze is fixed on a particular point, objects in the neighbourhood of that fixed point are seen as single (and not double) only when the line of sight makes the same angle to both eyes, as is the case for vision accommodated to infinity, and thus when the nearby object is imaged onto mutually corresponding points on the retina. The geometrical condition for this simple kind of vision, as it develops, is fulfilled precisely only on a particular three-dimensional curve, whose construction using the methods of descriptive geometry was treated by Grossmann.

Grossmann's last original research work again dealt with a question at the intersection of geometry research and practical applications [142]. The basics are again found to be described in one of his textbooks, this time in the last chapter of his "Descriptive Geometry for Mechanical Engineers". This problem deals with the geometrical description of a particular mechanism which is utilised in mechanical looms, and in which a rotational motion is converted into a periodic batting motion.

In mechanical looms, the weaving shuttle with the thread is thrown through the warp by the periodic batting motion of an arm. The batting arm is mounted at right angles on a shaft, and its periodic motion is produced by a second roller arm mounted below the first on the same shaft, which is forced out of its resting position by the motion of a cam. The cam wheel has a circular shape with an attached extension ["nose"] that presses against the roller arm at a certain position during its rotation. The cam wheel is manufactured in practice by using approximate templates and grinding as needed; there is no guarantee that the flank of the cam wheel has the proper shape so that it will press uniformly and smoothly against the roller arm. These shortcomings lead in practice to frictional losses and the necessity of frequent repairs. In one of his last works, Grossmann investigated the geometric situation during rotation of the cam wheel against the roller arm. He found that there is always a well-defined shape of the cam wheel which allows the surface of the wheel to follow a line of contact on the roller during its complete rotation. Considered geometrically, this corresponds to the rolling of curved regular surfaces of second order. With this insight into the geometric properties of the mechanical problem, Grossmann had the idea for a mechanical construction method which would permit grinding the cam wheel during its initial fabrication in such a way that it is geometrically optimised, that is it permits an optimal transfer of force and minimises frictional losses. He patented his idea in 1927 [143], and he received a certain amount of start capital from a foundation in order to realise his idea in practice. Unfortunately, it developed that his progressive illness did not allow him to develop his patent practically, and he returned a portion of the capital that he had been granted.

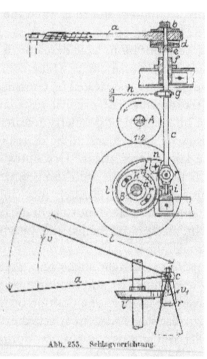

Abb. 255. Schlagvorrichtung.

Photo and figure The cam wheel and the roller arm of a mechanical loom. Grossmann analysed the geometric situation during rotation of the cam wheel against the roller arm, and patented a mechanism for the precise fabrication of cam wheels, whose flank would be in uniform contact, along a continuous curve, with the roller wheel at all points during its rotation. (Below the figure: *Fig. 255. Batting mechanism.*)

Grossmann's friendship with Einstein was both a blessing and a curse for him. Their close collaboration during their years as colleagues at the ETH led to two joint publications which treated a generalised theory of relativity, and these made him famous. As a result of this collaboration, Grossmann became one of the founders of General Relativity, and therewith of our modern understanding of space and time, of gravitation and cosmology. But this collaboration led Grossmann into a field which was not his actual area of specialisation, and it seems that he himself had perceived that. When Einstein wrote to Arnold Sommerfeld already in 1915 that "*Grossmann [will] never make a claim to being a codiscoverer*" [144], it is quite possible that he was only expressing what Grossmann himself had said to him. In any case, the fact that their close collaboration ended when Einstein left for Berlin speaks in favour of the idea that Grossmann may have seen his task as completed, or also that he had recognised his own limits in the matter.

This restraint changed once again after Grossmann's merciless illness had already taken possession of him. At the end of the 1920's, Einstein had begun to pursue his programme of a unified field theory, a unification of the two then-known fundamental forces of Nature, gravitation and electromagnetism, on the basis of General Relativity. Einstein was to continue this programme up to his death in 1955, without being able to find a solution. Today, most physicists see Einstein's efforts in his final years as unsuccessful, and are dubious that the path which he chose to follow could ever have led to reaching his goal. But this was not yet clear in the late 1920's. On the contrary, Einstein's approach to a unified field theory based on the so-called distant parallelism [or *teleparallelism*], which he pursued from 1928 to 1931, not only gave rise to great interest on the part of the media; it also appeared to be promising to many physicists and mathematicians at the time. Einstein's new approach in any case called up a number of mathematicians to action, who were familiar with and dealt with the mathematical fundamentals of Einstein's theory. The mathematics in question was an extension of the geometrical basis of General Relativity. Marcel Grossmann was one of those who became active.

Initially in private correspondence, then in an article in the "*Vierteljahrsschrift der Naturforschenden Gesellschaft*" [Quarterly of the Swiss Society of Natural Science Researchers], he attacked Einstein and criticised him for his—in Grossmann's opinion—incorrect understanding of the geometrical basis of distant parallelism. It is not surprising that Grossmann felt challenged by Einstein's new theory to make a statement—for Einstein had here introduced a geometrical theory into his considerations, and thus was treading on the field for which Grossmann felt responsible. And he was of the opinion that Einstein had not understood the basic geometrical concepts correctly here, and he believed in that sense that he should warn Einstein. In his publication, he wrote:

Already once before—it was in the year 1913—Einstein published 'field equations' found by this method, which had to be modified after only a few years; at that time, I shared the responsibility. [145]

And further on, he wrote: "*And also to the developments in differential geometry and mathematical physics of recent years, I say here 'no', from my conviction that I am thereby giving beneficial support to science and, in the final end, also to my friend.*" [146]

Even with a generous interpretation, Grossmann's refusal to accept Einstein's new theory, as well as the newer developments in differential geometry and mathematical physics, cannot be completely reconstructed in a rational manner. He even made elementary errors of computation in his arguments.

Here, we see the curse of his friendship with Einstein: Always being measured against the great achievements of his friend, in which he was also significantly involved during the time when the two of them were able to see each other daily and discuss their work. But the limelight is only one side of his life's work; his cultivation of mathematics in general and of geometry in particular, in both teaching and research; his function as an academic teacher at the ETH over twenty years until his early retirement, forced upon him by an insidious illness; his commitment as an educator and organiser in many areas, all these deserve a regard which is not to be overshadowed by his friendship with Einstein.

Backnotes

1. Memoirs of the life of Jules Grossmann, dictated to his granddaughter Elsbeth Grossmann, 1932.
2. *"Die Schweizerkolonie in Ungarn 1867–1990"* ["The Swiss Colony in Hungary, 1867–1990"]: Dissertation by Antal András Kováts, University of Hagen 2012, pp. 80, 88.
3. *"Die Schweizerkolonie in Ungarn 1867–1990"* ["The Swiss Colony in Hungary, 1867–1990"]: Dissertation by Antal András Kováts, University of Hagen 2012, p. 48.
4. *"Wien, Prag, Budapest: Blütezeit der Habsburgermetropolen"* ["Vienna, Prague, Budapest: the Heyday of the Hapsburg Metropolises"], 1996, *Promedia Druck- und Verlagsgesellschaft m.b.H.*, Vienna, p. 61.
5. *"288. Neujahrsblatt der Stadt Winterthur"* [288th New Year's Folio of the City of Winterthur], by Emil Wegmann, 1957.
6. Memoirs of Eugen Grossmann, 1959.
7. *"Wien, Prag, Budapest: Blütezeit der Habsburgermetropolen"* ["Vienna, Prague, Budapest: the Heyday of the Hapsburg Metropolises"], 1996, *Promedia Druck- und Verlagsgesellschaft m.b.H.*, Vienna, p. 66.
8. *"Vergessene 'Judenhäuser' in Budapest"* ["Forgotton 'Jewish Houses' in Budapest"], by Meret Baumann, *Neue Zürcher Zeitung* (NZZ), June 22nd, 2014.
9. www.wiener-weltausstellung.at.
10. *Aargauer Zeitung*, November 11th, 2013.
11. *"Die Chronik derer zu Lichtenhayn"* ["The Chronicle of the Lichtenhayn Family"], by René Falconnier, 1979.
12. *curriculum vitae* of Ernest Grossmann, 1892.

© Springer International Publishing AG, part of Springer Nature 2018
C. Graf-Grossmann, *Marcel Grossmann*, Springer Biographies,
https://doi.org/10.1007/978-3-319-90077-3

13. *Stadtarchiv Schaffhausen* [Municipal Archives of Schaffhausen], C II 10.01.01/09.

14. *"Familie Schulthess von Zürich"* ["The Schulthess Family of Zurich"], *Festschrift* by Emil Usteri, 1958, Schulthess & Co. AG, Zurich, p. 93.

15. *Stadtarchiv Schaffhausen* [Municipal Archives of Schaffhausen], D IV 01.06.07/11. (Balance sheet for 1889/90)

16. *Stadtarchiv Schaffhausen* [Municipal Archives of Schaffhausen], D IV 01.06.07/11. (Minutes of the General Meeting, 1890)

17. *"Familie Schulthess von Zürich"* ["The Schulthess Family of Zurich"], *Festschrift* by Emil Usteri, 1958, Schulthess & Co. AG, Zurich, p. 95.

18. Swiss Federal Office of Statistics.

19. *"Zürich, La Belle Epoque"* ["Zurich, the *Belle Epoque*"], by Walter Baumann, 1973, p. 14.

20. *"Alfred Escher: Der Aufbruch zur modernen Schweiz"* ["Alfred Escher: The Dawn of Modern Switzerland"], by Joseph Jung, 2006, Vol. 2, pp. 855–857.

21. *Quartals-Zeugnis der Oberen Realschule zu Basel, ETH-Archiv Zürich, Hs* 421a:17 [Quarterly Report Cards of the Higher Middle School in Basel], Archives of the ETH Zurich.

22. *StABS Erziehung T 9b_IV A, Herbst 1896, Stadtarchiv Basel-Stadt, Archivsignatur* T 9b [Municipal Archives of the City of Basel, Autumn 1896, StABS 'Education', Call number T 9b].

23. Notes by Eugen Grossmann, ca. 1900.

24. "Marcel Grossmann and his Contribution to the General Theory of Relativity", Tilman Sauer, 2014; see Note 127.

25. *"Zürich, La Belle Epoque"* ["Zurich, the *Belle Epoque*"], by Walter Baumann, 1973, p. 85.

26. *"Nachruf Marcel Grossmann"* ["Epitaph for Marcel Grossmann"], by Louis Kollros, extracted from the files of the *Neue Helvetische Gesellschaft*, Geneva, 1937. Original text: *"J'ai eu le privilège d'être alors son camarade d'études. Il m'avait immédiatement frappé par sa vivacité et son entrain; il comprenait tout avec une rapidité surprenante; le travail était pour lui un jeu; il nous animait tous par sa gaité, son esprit critique et sa verve pittoresque; il savait découvrir les petits travers, mais aussi les qualités de ses camarades et de ses maîtres."*

27. Letter from Albert Einstein to Marcel Grossmann, April 14th, 1901, in The Collected Papers of Albert Einstein (CPAE), Vol. 1, 1987, English translation supplement, Doc. 100; Original in the Archives of the ETH Zurich, Hs 412a:1. See http://einsteinpapers.press.princeton.edu/vol1-trans/.

28. www.archiv.uzh.ch/editionen/uzh_dozierendenverzeichnis_1833–1933.pdf.

29. CPAE, Vol. 1, 1986, Doc. 122; Original in the Estate of Marcel Grossmann, *Bernisches Historisches Museum* (BHM), Inventory no. 62778. Online: cf. Note 27.

30. "*Spiessrutenlauf für ein Genie*" ["Running the Gauntlet for a Genius"], by Alexis Schwarzenbach, *Weltwoche* 20/2004.

31. Memoirs of Marcel Grossmann Jr., 1983.

32. Postcard from Albert Einstein to Marcel Grossmann, April 6th, 1904. CPAE, Vol. 5, 1993, Doc. 17. Original in the private possession of the Grossmann family. Online: cf. Note 27.

33. "*Eine neue Bestimmung der Moleküldimensionen*" ["A New Determination of Molecular Dimensions"], Albert Einstein, *Annalen der Physik*, 4th Series, 1900–1928, Vol. 1–87 (1905).

34. Max-Planck-Institut für Gravitationsphysik, Einstein Online, www.einstein-online.info.

35. *loq. cit.*, Einstein Online, www.einstein-online.info.

36. *Archiv Einwohnerkontrolle* [Census Archives], Community of Basel, Municipal Archives of the City of Basel, PD-REG 14a 85 36142.

37. Minutes of the Faculty meetings, *StABS*, University Archives R 6.3.

38. ETH Library, Archives, SR2: *Präsidialverfügungen* [Presidential Enactments] 1906, Enactment No. 456, dated October 15th, 1906 and No. 532, dated November 23rd, 1906.

39. Letter from Albert Einstein to Marcel Grossmann, November 18th, 1911. CPAE, Vol. 5, 1993, Doc. 307. Original in the Estate of Marcel Grossmann, *Bernisches Historisches Museum* (BHM), Inventory no. 62782. Online: cf. Note 27.

40. Letter from Albert Einstein to Marcel Grossmann, December 12th, 1912, CPAE, Vol. 5, 1993, Doc. 319; Original in the Archives of the ETH Zürich, Hs 421a:6. Online: cf. Note 27.

41. Schools Inspector Minutes of the Meeting on December 2nd, 1911, No. 17, ETH Archives.

42. "Mathematicians and Mathematics in Zurich, at the University and at ETH", by Urs Stammbach and Gunther Frei, *Schriftenreihe A* [Publications Series A] of the ETH Library, Vol. 8, 2007.

43. The history of General Relativity has been described by a number of analogies and metaphors. John Stachel, for example, describes it as a play in three acts.

44. "*Nachruf Marcel Grossmann*" ["Epitaph for Marcel Grossmann"], by Louis Kollros, extracted from the files of the *Neue Helvetische*

Gesellschaft, Geneva, 1937. Original text: "*En 1913, il est pris dans un engrenage scientifique qui lui procure à la fois un travail intense et un très grand plaisir. Son ami et camarade d'étude Albert Einstein avait déjà créé sans utiliser de mathématiques supérieures ce qu'on appelle aujourd'hui la Relativité restreinte. Mais quand il fut amené à sa Relativité généralisée et à sa théorie de la gravitation, il fut acculé à des difficultés mathématiques telles qu'il vint un jour consulter son ami. Marcel Grossmann sut lui montrer que l'instrument mathématique indispensable à l'édification de la nouvelle physique avait été trouvé en 1869 à Zurich par Christoffel, le créateur et le premier doyen de la section autonome de mathématiques et physique à l'Ecole polytechnique fédérale.*"

45.   Letter, Albert Einstein to Arnold Sommerfeld. CPAE, Vol. 5, 1993, Doc. 421. Online: cf. Note 27.

46.   "*Kovarianzeigenschaften der Feldgleichungen der auf die verallgemeinerte Relativitatstheorie gegründeten Gravitationstheorie*" ["Covariance Properties of the Field Equations of a Theory of Gravitation based upon the Generalised Theory of Relativity"], by Albert Einstein and Marcel Grossmann, *Zeitschrift für Mathematik und Physik*, May 29th, 1914. CPAE, Vol. 6, Doc. 2. Online: cf. Note 27.

47.   *Sitzungsberichte, 1030, Königliche Preußische Akademie der Wissenschaften* [Proceedings of the Royal Prussian Academy of Sciences], 1030–1085, Albert Einstein, 1914.

48.   CPAE, Vol. 8, 1998, Doc. 75. Online: cf. Note 27.

49.   CPAE, Vol. 8 1998, Doc. 2. Online: cf. Note 27.

50.   ETH Library, Zurich, Hs 421:2.

51.   "*Nachruf Marcel Grossmann*" ["Epitaph for Marcel Grossmann"], by Louis Kollros, extracted from the files of the *Neue Helvetische Gesellschaft*, Geneva, 1937. Original text: "*La première impression qu'il faisait sur ses étudiants était celle d'un chef et d'un entraîneur; ses cours étaient d'une merveilleuse clarté; il avait le don d'enthousiasmer ses auditeurs. [...] Grossmann s'est efforcé de trouver des exercices intéressants, de proposer à ses élèves de jolies applications de la théorie. [...] Il suivait ses élèves après leurs études et se donnait beaucoup de peine pour leur procurer une place. Chacun désirait être assistant chez lui.*"

52.   www.odeon.ch.

53.   "*Darstellende Geometrie in systematischen Beispielen*" ["Descriptive Geometry in Systematic Examples"], Klaus Ulshofer and Dietrich Tilp, *Verlag C. C. Buchner*, 2011.

54.   ETH Library Zurich, University Archives, *Darstellende Geometrie* [Descriptive Geometry], Hs 421:35.

55. *"Diagnosen"* [Diagnoses], Essay, Marcel Grossmann, 1926.

56. www.math.ch: *"100 Jahre SMG"*, [100 Years of the SMG], 2010.

57. Eugen Grossmann, *Historisches Lexikon der Schweiz* [Historical Encyclopedia of Switzerland], www.hls-dhs-dss.ch.

58. *"Insel der unsicheren Geborgenheit"* ["The Island of Uncertain Security"], by Georg Kreis, *Verlag Neue Zürcher Zeitung*, 2014.

59. *"14/18, der Weg nach Versailles"* ["1914/1918, the Path to Versailles"], by Jörg Friedrich, *Ullstein Buchverlag GmbH*, 2014.

60. *"Kritischer Patriotismus"* ["Critical Patriotism"], by Catherine Guanzini and Peter Wegelin, *Verlag Paul Haupt*, 1989.

61. *"Les Helvetistes"* ["The Helvetians"], by Alain Clavien, *Societé d'Histoire de la Suisse Romande & Editions d'En Bas*, 1993, pp. 269 ff.

62. *"Les Helvetistes"* ["The Helvetians"], by Alain Clavien, *Societé d'Histoire de la Suisse Romande & Editions d'En Bas*, 1993, p. 285.

63. *"Kritischer Patriotismus"* ["Critical Patriotism"], by Catherine Guanzini and Peter Wegelin, *Verlag Paul Haupt*, 1989, p. 37.

64. *"Warum wir auf Deutschlands Seite standen"* ["Why We Were on Germany's Side"], Thomas Maissen, *NZZ am Sonntag*, June 29th, 2014.

65. *"Insel der unsicheren Geborgenheit"* ["The Island of Uncertain Security"], by Georg Kreis, *Verlag Neue Zürcher Zeitung*, 2014, p. 205.

66. *"Les Helvetistes"* ["The Helvetians"], by Alain Clavien, *Societé d'Histoire de la Suisse Romande & Editions d'En Bas*, 1993, pp. 40ff.

67. *"Notizen eines Müssiggängers"* ["Notes from an Idler"], by J. R. von Salis, *Orell Fussli Verlag*, 1983, p. 176.

68. Website of the NHG-TS: www.dialoguesuisse.ch.

69. *"Sinn und Tragweite der eidgenössischen Maturitätsreform"* [The Purpose and Scope of the Federal School-Leaving Certificate Reform], Reprint from the NZZ, 1925.

70. *Festschrift* for the 80th birthday of Marcel Grossmann, Max and Liesel Keller, 1984.

71. www.schauspielhaus.ch.

72. Information from the Notary of Zurich-Hottingen, obtained on July 7th 2014.

73. Quoted from the Memoirs of Henriette Hogerzeil-Grossmann, 1987. Original text: *"The idea was well meant, but it didn't come out so well. As much as the two brothers loved each other, the two sisters had nothing at all in common. My mother was and remained a very spoilt and selfish girl, had a very comfortable and easy life, and wanted always to play the first role."*

74. ETHeritage, Highlights from the Collections and Archives of the ETH Library. Marion Wullschleger, Blog entry on August 8th, 2014.

75. Article in the NZZ, first morning edition on August 13th, 1915.

76. *"Nachruf Marcel Grossmann"* ["Epitaph for Marcel Grossmann"], by Louis Kollros, extracted from the files of the *Neue Helvetische Gesellschaft*, Geneva, 1937. Original text: *"… il créa avec le professeur Egger de l'Université de Zurich son journal courageux et indépendant: la 'Neue Schweizer Zeitung' …"*

77. *"Insel der unsicheren Geborgenheit"* ["The Island of Uncertain Security"], by Georg Kreis, *Verlag Neue Zürcher Zeitung*, 2014, pp. 72ff.

78. State Archives of the Canton of Zürich, Call number SPS-NSZ, Z 2.838.

79. Article from the *Neue Schweizer Zeitung*, 1919 – 1922.

80. Article by Albert Egger, *Neue Schweizer Zeitung*, June 1922.

81. CPAE, Vol. 10, 2006, Introduction. Online: cf. Note 27; here: Main volume, Introduction, Sect. V.

82. Letter from Marcel Grossmann to Albert Einstein, September 9th, 1920. CPAE, Vol. 10, 2006, Doc. 142. Original in the Einstein Archive, No. 11-461. Online: cf. Note 27.

83. Letter from Albert Einstein to Marcel Grossmann, September 12th, 1920. CPAE, Vol. 10, 2006, Doc. 148. Original in private possession of the Grossmann family. Online: cf. Note 27.

84. CPAE, Vol. 10, 2006, Doc. 148. Online: cf. Note 27.

85. CPAE, Vol. 10, 2006, Doc. 241. Online: cf. Note 27; however only in German (not selected for translation).

86. Postcard from Albert Einstein to Marcel Grossmann, October 7th, 1920. CPAE, Vol. 13, 2012, Doc. 382. Online: cf. Note 27.

87. Postcard from Albert Einstein to Marcel Grossmann, November 23rd, 1922. CPAE, Vol. 14, 2015, (Vol. 13), Doc. 386a. Original at the BHM, Inventory no. 62783. Note: 'Fucisava' = R. Fujisawa, a Japanese mathematician whom Marcel had met in Cambridge in 1912. Online: cf. Note 27; however only in German (not selected for translation).

88. Letter from Marcel Grossmann to Albert Einstein, August 1st, 1923. CPAE, Vol. 14, Doc. 97. Original in the Einstein Archive, no. II-464. Online: cf. Note 27.

89. Letter from Albert Einstein to Marcel Grossmann, December 28th, 1923. CPAE, Vol. 14, 2015, Doc. 186. Original in the BHM, Inventory no. 62784. Online: cf. Note 27.

90. Letter from Marcel Grossmann to Albert Einstein, January 11th, 1924. Original in the Einstein Archive, no. II-469.

91. Postcard from Albert Einstein to Marcel Grossmann, March 15th, 1924. CPAE, Vol. 14, 2015, Doc. 226. Original at the BHM, Inventory no. 62779. Online: cf. Note 27.

92. Marcel Grossmann to Heinrich Zangger, March 12th, 1927. Einstein Archive, no. 40059.

93. Recollections of Elsbeth Grossmann of her father's illness, transcribed by Heidi Grossmann on November 24th, 1982.

94. See www.wikipedia.org on Emile Coué.

95. *"Seelenverwandte"* ["Soulmate"], by Robert Schulman, *Verlag Neue Zürcher Zeitung*, 2012, p. 417f.

96. *"Seelenverwandte"* ["Soulmate"], by Robert Schulman, *Verlag Neue Zürcher Zeitung*, 2012, p. 443f.

97. *"Seelenverwandte"* ["Soulmate"], by Robert Schulman, *Verlag Neue Zürcher Zeitung*, 2012, p. 445.

98. *"Seelenverwandte"* ["Soulmate"], by Robert Schulman, *Verlag Neue Zürcher Zeitung*, 2012, p. 446.

99. www.ms-diagnose.ch.

100. Patent CH121538(A), GB286710(A), State Archives of the Canton of Zürich, Call number PAT2, 21C, No. 121538.

101. Diaries of Marcel Hans Grossmann, condensed, translated into High German and transcribed by Heidi Grossmann, 1989.

102. Funeral speech in memory of Mr. Jules Grossmann-Lichtenhahn, by Peter Marti, Thalwil, 1934.

103. Rüdiger Vaas, in *"Bild der Wissenschaft"*, November 2014.

104. www.wissenschaft-online.de.

105. Letter from Marcel Grossmann to Albert Einstein, November 11th, 1930. Einstein Archive, no. 11-475.

106. Letter from Marcel Grossmann to Albert Einstein, April 1931. Einstein Archive, no. 11-478.

107. Letter from Albert Einstein to Marcel Grossmann, June 9th, 1931. In the private possession of the Grossmann family.

108. Letter from Marcel Grossmann to Albert Einstein, June 23rd, 1931. Einstein Archive, no. 11-480.

109. Quarterly of the Natural Science Research Society in Zurich, 1936, pp. 324f.

110. *Neue Zürcher Zeitung* (NZZ), September 13th, 1936.

111. Albert Einstein to Anna Grossmann, September 20th, 1936; in: *"Albert Einstein und die Schweiz"* ["Albert Einstein and Switzerland"], by Carl Seelig, Europa-Verlag, 1952, p. 177. Original in the private possession of the Grossmann family.

112. On his School Leaving Certificate, Grossmann received only the highest marks of 6, except in English and Technical Drawing, for which he received a 5. He was thus, together with one other pupil, the best in his class of 18 graduates (cf. Note 22).

113. From his graduating class, only one other of his fellow graduates planned to go to the Polytechnic, and his planned major subject was given as "undetermined".

114. ETH Library, *Hochschularchiv* [University Archives], Hs 421:2.

115. 115. Grossmann, Marcel (1904): "*Die fundamentalen Konstruktionen der nichteuklidischen Geometrie*" ["The Fundamental Constructions of Non-Euclidean Geometry"]. *Beilage zum Programm der Thurgauischen Kantonsschule für das Schuljahr 1903/04*, Frauenfeld: Huber & Co.

116. Grossmann, Marcel (1904): "*Die Konstruktion des geradlinigen Dreiecks der nicht-euklidischen Geometrie aus den drei Winkeln*" ["The Construction of the Linear Triangle in Non-Euclidean Geometry from its Three Angles"], Mathematische Annalen **58**, pp. 578–582.

117. Bonola, Roberto (1908): "*Die Nichteuklidische Geometrie, Historisch-kritische Darstellung ihrer Entwicklung*" ["Non-Euclidean Geometry, A Historical-Critical Account of its Development"], Leipzig: Teubner, pp. 181, 240.

118. Liebmann, Heinrich (1911): "*Die elementaren Konstruktionen der nichteuklidischen Geometrie*" ["The elementary Constructions of Non-Euclidean Geometry"], *Jahresbericht der Deutschen Mathematiker-Vereinigung* **20**, pp. 58–69, here p. 69.

119. Zacharias, Max (1913): "*Elementargeometrie und elementare nicht-euklidische Geometrie in synthetischer Behandlung*" ["Elementary Geometry and Elementary Non-Euclidean Geometry: A Synthetic Treatment"], in: *Encyclopädie der mathematischen Wissenschaften mit Einschluss ihrer Anwendungen* [Encyclopedia of the Mathematical Sciences, Including their Applications], Vol. 3: *Geometrie* (1914–1931). *Erster Theil, zweite Hälfte, Nr. 9*, pp. 859–1172, here p. 1159.

120. Construction problems in non-Euclidean hyperbolic geometry were later the topics of the four dissertations completed under Grossmann's direction: Ernst Vaterlaus, "*Konstruktionen in der Bildebene der hyperbolischen Zentralprojektion*", 1916; Ernst Mettler, "*Anwendung der stereographischen Projektion und Konstruktionen im nicht-euklidischen Raume*", 1916; Karl Dändliker, "*Darstellende hyperbolische Geometrie*", 1919; Emil Leutenegger, "*Über Kegelschnitte in der hyperbolischen Geometrie*", 1923.

121. Grossmann, Marcel (1906): *"Darstellende Geometrie. Leitfaden für den Unterricht an höheren Lehranstalten"* ["Descriptive Geometry. A Guiding Textbook for Instruction at Higher Schools"], Basel: Helbing & Lichtenhahn; Grossmann, Marcel (1906): *"Analytische Geometrie. Leitfaden für den Unterricht an höheren Lehranstalten"* ["Analytic Geometry. A Guiding Textbook for Instruction at Higher Schools"], Basel: Helbing & Lichtenhahn.

122. Fiedler, Otto Wilhelm (1883–1888): *"Die darstellende Geometrie in organischer Verbindung mit der Geometrie der Lage"* ["Descriptive Geometry in a Systematic Combination with Positional Geometry"]. *I. Theil. Die Methoden der darstellenden und die Elemente der projectivischen Geometrie; II. Theil. Die darstellende Geometrie der krummen Linien und Flächen; III. Theil. Die constituirende und analytische Geometrie der Lage*, Leipzig: B.G. Teubner.

123. Salmon, George (1898): *"Analytische Geometrie des Raumes"* ["Analytic Geometry of Space"] (German edited by W. Fiedler). *I. Theil. Die Elemente und die Theorie der Flächen zweiten Grades; II. Theil. Analytische Geometrie der Curven im Raume und der algebraischen Flächen*, Leipzig: Teubner.

124. See: Geiser, Carl Friedrich (1921): *"Zur Erinnerung an Theodor Reye"* ["In Memory of Theodor Reye"], *Naturforschende Gesellschaft Zürich. Vierteljahrsschrift* (Vol. 66), [Quarterly of the Natural Science Research Society in Zurich], pp. 158–180. For instruction in French, Louis Kollros wrote a similar treatment of *"Géométrie déscriptive"* (Zurich: Orell Füssli, 1918).

125. Grossmann, Marcel (1913): *"Prof. Dr. Otto Wilhelm Fiedler (1832–1912)"* [Epitaph for Prof. Otto Wilhelm Fiedler], *Schweizerische Naturforschende Gesellschaft. Verhandlungen* (Vol. 96) [Proceedings of the Swiss Society of Natural Science Research], pp. 20–27.

126. Grossmann, Marcel (1911): *"Der mathematische Unterricht an der Eidgenössischen Technischen Hochschule"* [Mathematics Instruction at the ETH Zürich], Basel and Geneva: Georg & Co.

127. For a detailed treatment, see Sauer, Tilman (2015): *"Marcel Grossmann and his contribution to the general theory of relativity"*, in Proceedings of the 13th Marcel Grossmann Meeting on Recent Developments in Theoretical and Experimental General Relativity, Gravitation, and Relativistic Field Theory, 2012. Edited by R.T. Jantzen, K. Rosquist, R. Ruffini. Singapore: World Scientific, 2015, pp. 456–503 [arXiv:1312.4068].

128. Christoffel, E.B. (1869): *"Über die Transformation der homogenen Differentialausdrücke zweiten Grades"* ["On the Transformation of Homogeneous Differential Expressions of Second Order"], *Journal für die reine und angewandte Mathematik* **70**, pp. 46–70.

129. Riemann, Bernhard (1876): *"Commentatio mathematica, qua respondere tentatur quaestioni ab Illma Academia Parisieni propositae"* [Mathematical comments, which attempt to answer questions proposed by the Illustrous Parisian Academy], in: *Gesammelte Mathematische Werke und wissenschaftlicher Nachlass* [Collected Mathematical Works and Scientific Heritage], pp. 370–383.

130. Bianchi, Luigi (1899): *"Vorlesungen über Differentialgeometrie" (autorisierte deutsche Übersetzung von Max Lukat)* [Lectures on Differential Geometry (authorised German translation by Max Lukat)], Leipzig: Teubner.

131. Ricci, Gregorio und Levi-Civita, Tullio (1901): *"Méthodes de calcul différentiel absolu et leurs applications"* [Methods of Absolute Differential Calculus and their Applications], *Mathematische Annalen* **54**, pp. 125–201.

132. For a detailed reconstruction and interpretation of this notebook, see: J. Renn (Ed.) *"The Genesis of General Relativity"*. Vol. 1: Einstein's Zurich Notebook. Introduction and Source. Vol. 2: Einstein's Zurich Notebook. Commentary and Essays, Dordrecht: Springer, 2007.

133. For a more detailed discussion of the 'Outline', see the work cited in the previous Note, as well as (among others): Pais, A.: " *'Subtle is the Lord …' The Science and the Life of Albert Einstein"*, Oxford: Oxford University Press, 1982, Chap. 12; Norton, J.D. (1984): *"How Einstein Found His Field Equations"*. Historical Studies in the Physical Sciences, **14**, pp. 253–316; Stachel, J.: *"Einstein from 'B' to 'Z'"*, Boston: Birkhäuser, 2002, Chap. V; Sauer, T. (2005): *"Einstein's Review Paper on General Relativity"*, in: Landmark Writings of Western Mathematics, 1640–1940, I. Grattan-Guinness (Ed.), Amsterdam: Elsevier, 2005, pp. 802–822; Janssen, M. (2014): " *'No Success Like Failure…' Einstein's Quest for General Relativity, 1907–1920"*. In "The Cambridge Companion to Einstein", M. Janssen & C. Lehner (Eds.), Cambridge: Cambridge University Press, pp. 167–227; Gutfreund, H., and Renn. J.: *"The Road to Relativity"*, Princeton: Princeton University Press, 2015; as well as the other literature cited in these works.

134. Einstein, Albert, und Grossmann, Marcel (1913): *"Entwurf einer verallgemeinerten Relativitätstheorie und einer Theorie der Gravitation"* [Outline of a Generalised Theory of Relativity and of a Theory

of Gravitation], *Zeitschrift für Mathematik und Physik* **62**, No. 3, pp. 225–261, here p. 245.

135. Einstein, Albert, und Grossmann, Marcel (1914): "*Kovarianzeigenschaften der Feldgleichungen der auf die verallgemeinerte Relativitätstheorie gegründeten Gravitationstheorie*" [Covariance Properties of the Field Equations of the Theory of Gravitation based upon the Generalised Theory of Relativity], *Zeitschrift für Mathematik und Physik* **63**, pp. 215–225. (See also Backnote [46].)

136. Einstein, Albert (1915): "*Zur allgemeinen Relativitätstheorie*" [On the General Theory of Relativity], *Königlich Preußische Akademie der Wissenschaften* [Proceedings of the Royal Prussian Academy of Sciences], *Sitzungsberichte* (1915), pp. 778–786, here p. 778.

137. Grossmann, Marcel (1927): "*Darstellende Geometrie für Maschineningenieure*" [Descriptive Geometry for Mechanical Engineers], Berlin: Springer, p. IV.

138. See in particular Chapter 10 of this biography.

139. Grossmann, Marcel (1910): "*Lösung eines geometrischen Problems der Photogrammetrie*". [Solution of a Geometric Problem of Photogrammetry]. *Schweizerische Naturforschende Gesellschaft. Verhandlungen*, [Proceedings of the Swiss Society for Natural Science Research] **93**, pp. 338–339.

140. Grossmann, Marcel (1925): "*Das vollständige Fokalsystem einer ebenen algebraischen Kurve*". [The Complete Focal System of a Planar Algebraic Curve]. *Acta Litterarum ac Scientiarum* (Szeged). *Sectio Scientarum Mathematicarum*, **2**, pp. 178–181.

141. Grossmann, Marcel (1925): "*Darstellung des Horopters*". [Representation of the Horopter Curve]. *Naturforschende Gesellschaft Zürich. Vierteljahrsschrift* [Quarterly of the Natural Science Research Society in Zurich], **70**, pp. 66–76.

142. Grossmann, Marcel (1927): "*Präzisions-Schlagexzenter für mechanische Webstühle: geometrische Formgebung und zwangläufige Herstellung*" [A Precision Camwheel for Mechanical Looms: Geometrical Form and Exact Fabrication]. *Schweizerische Bauzeitung* [Swiss Construction Journal], **90**, pp. 279–282.

143. Patent CH121538(A), GB286710(A).

144. Albert Einstein to Arnold Sommerfeld, July 15th, 1915, in: The Collected Papers of Albert Einstein (CPAE), Vol. 8: The Berlin Years, Correspondence, 1914–1918, Princeton, Princeton University Press, 1998, Doc. 96. Online: cf. Note 27.

145. Grossmann, Marcel (1931): *"Fernparallelismus? Richtigstellung der gewählten Grundlagen für eine einheitliche Feldtheorie"* [Teleparallelism? Correction of the Fundamentals Chosen for a Unified Field Theory]. *Naturforschende Gesellschaft Zürich. Vierteljahrsschrift* [Quarterly of the Natural Science Research Society in Zurich], **76**, pp. 42–60, here p. 54.

146. Grossmann, Marcel (1931): *op. cit.*, Note 145; here p. 59.

ENTWURF EINER
VERALLGEMEINERTEN RELATIVITÄTSTHEORIE
UND EINER
THEORIE DER GRAVITATION

I. PHYSIKALISCHER TEIL
VON
ALBERT EINSTEIN
IN ZÜRICH

II. MATHEMATISCHER TEIL
VON
MARCEL GROSSMANN
IN ZÜRICH

LEIPZIG UND BERLIN

DRUCK UND VERLAG VON B. G. TEUBNER

1913

Separatabdruck aus
„Zeitschrift für Mathematik und Physik" Band 62.

I.

Physikalischer Teil.

Von Albert Einstein.

Die im folgenden dargelegte Theorie ist aus der Überzeugung hervorgegangen, daß die Proportionalität zwischen der trägen und der schweren Masse der Körper ein exakt gültiges Naturgesetz sei, das bereits in dem Fundamente der theoretischen Physik einen Ausdruck finden müsse. Schon in einigen früheren Arbeiten[1]) suchte ich dieser Überzeugung dadurch Ausdruck zu verleihen, daß ich die schwere auf die träge Masse zurückzuführen suchte; dieses Bestreben führte mich zu der Hypothese, daß ein (unendlich wenig ausgedehntes homogenes) Schwerefeld sich durch einen Beschleunigungszustand des Bezugssystems physikalisch vollkommen ersetzen lasse. Anschaulich läßt sich diese Hypothese so aussprechen: Ein in einem Kasten eingeschlossener Beobachter kann auf keine Weise entscheiden, ob der Kasten sich ruhend in einem statischen Gravitationsfelde befindet, oder ob sich der Kasten in einem von Gravitationsfeldern freien Raume in beschleunigter Bewegung befindet, die durch an dem Kasten angreifende Kräfte aufrecht erhalten wird (Äquivalenz-Hypothese).

Daß das Gesetz der Proportionalität der trägen und der schweren Masse jedenfalls mit außerordentlicher Genauigkeit erfüllt ist, wissen wir aus einer fundamental wichtigen Untersuchung von Eötvös[2]), die auf folgender Überlegung beruht. Auf einen an der Erdoberfläche ruhenden Körper wirkt sowohl die Schwere als auch die von der Drehung der Erde herrührende Zentrifugalkraft. Die erste dieser Kräfte ist proportional der schweren, die zweite der trägen Masse. Die Richtung der Resultierenden dieser beiden Kräfte, d. h. die Richtung der scheinbaren Schwerkraft (Lotrichtung) müßte also von der physikalischen Natur des ins Auge gefaßten Körpers abhängen, falls die Proportionalität der trägen und schweren Masse nicht erfüllt wäre. Es ließen sich dann die scheinbaren Schwerkräfte, welche auf Teile eines heterogenen starren Systems wirken, im allgemeinen nicht zu einer Resultierenden vereinigen; es bliebe vielmehr im allgemeinen ein Drehmoment der scheinbaren

---

1) A. Einstein, Ann. d. Physik 4. 35. S. 898; 4. 38. S. 355; 4. 38. S. 443.

2) B. Eötvös, Mathematische und naturwissenschaftliche Berichte aus Ungarn VIII 1890. Wiedemann, Beiblätter XV. S. 688 (1891).

Schwerkräfte übrig, das sich beim Aufhängen des Systems an einem torsionsfreien Faden hätte bemerkbar machen müssen. Indem Eötvös die Abwesenheit solcher Drehmomente mit großer Sorgfalt feststellte, bewies er, daß das Verhältnis beider Massen für die von ihm untersuchten Körper mit solcher Genauigkeit von der Natur des Körpers unabhängig war, daß die relativen Unterschiede die dies Verhältnis von Stoff zu Stoff noch besitzen könnte, kleiner als ein Zwanzigmilliontel sein müßte.

Beim Zerfall radioaktiver Stoffe werden so bedeutende Energiemengen abgegeben, daß die Änderung der trägen Masse des Systems, welche nach der Relativitätstheorie jener Energieabnahme entspricht, gegenüber der Gesamtmasse nicht sehr klein ist.[1] Beim Zerfall von Radium beträgt z. B. jene Abnahme $\frac{1}{10\,000}$ der Gesamtmasse. Würden jenen Änderungen der trägen Masse nicht Änderungen der schweren Masse entsprechen, so müßten Abweichungen der trägen von der schweren Masse bestehen, die weit größer sind, als es die Eötvösschen Versuche zulassen. Es muß also als sehr wahrscheinlich betrachtet werden, daß die Identität der trägen und der schweren Masse exakt erfüllt ist. Aus diesen Gründen scheint mir auch die Äquivalenzhypothese, welche die physikalische Wesengleichheit der schweren mit der trägen Masse ausspricht, einen hohen Grad von Wahrscheinlichkeit zu besitzen.[2]

§ 1. Bewegungsgleichungen des materiellen Punktes im statischen Schwerefeld.

Gemäß der gewöhnlichen Relativitätstheorie[3] bewegt sich ein kräftefrei bewegter Punkt nach der Gleichung

$$(1) \qquad \delta\left\{\int ds\right\} = \delta\left\{\int \sqrt{-dx^2 - dy^2 - dz^2 + c^2 dt^2}\right\} = 0.$$

Denn es besagt diese Gleichung nichts anderes, als daß sich der materielle Punkt geradlinig und gleichförmig bewegt. Es ist dies die Bewegungsgleichung in Form des Hamiltonschen Prinzipes; denn wir können auch setzen

$$(1\,\text{a}) \qquad \delta\left\{\int H dt\right\} = 0,$$

wobei

$$H = -\frac{ds}{dt}\,m$$

---

[1] Die Abnahme der trägen Masse, die der abgegebenen Energie E entspricht, ist bekanntlich $\frac{E}{c^2}$, wenn mit c die Lichtgeschwindigkeit bezeichnet wird.

[2] Vgl. auch § 7 dieser Arbeit.

[3] Vgl. M. Planck, Verh. d. deutsch. phys. Ges. 1906. S. 136.

gesetzt ist, falls m die Ruhemasse des materiellen Punktes bedeutet. Hieraus ergeben sich in bekannter Weise Impuls J_x, J_y, J_z und Energie E des bewegten Punktes:

$$(2) \quad \begin{cases} J_x = m\,\dfrac{\partial H}{\partial \dot{x}} = m\,\dfrac{\dot{x}}{\sqrt{c^2 - q^2}}\,; \text{ etc} \\[3mm] E = \dfrac{\partial H}{\partial \dot{x}}\,\dot{x} + \dfrac{\partial H}{\partial \dot{y}}\,\dot{y} + \dfrac{\partial H}{\partial \dot{z}}\,\dot{z} - H = m\,\dfrac{c^2}{\sqrt{c^2 - q^2}}\,. \end{cases}$$

Diese Darstellungsweise unterscheidet sich von der üblichen nur dadurch, daß in letzterer J_x, J_y, J_z und E noch einen Faktor c aufweisen. Da aber c in der gewöhnlichen Relativitätstheorie konstant ist, so ist das hier gegebene System dem gewöhnlich gegebenen äquivalent. Der einzige Unterschied ist der, daß J und E andere Dimensionen besitzen als in der üblichen Darstellungsweise.

In früheren Arbeiten habe ich gezeigt, daß die Äquivalenzhypothese zu der Folgerung führt, daß in einem statischen Gravitationsfelde die Lichtgeschwindigkeit c vom Gravitationspotential abhängt. Ich gelangte so zu der Meinung, daß die gewöhnliche Relativitätstheorie nur eine Annäherung an die Wirklichkeit gebe; sie sollte in dem Grenzfalle gelten, daß in dem betrachteten Raum-Zeitgebiete keine zu große Verschiedenheiten des Gravitationspotentials auftreten. Außerdem fand ich als Gleichungen der Bewegung eines Massenpunktes in einem statischen Gravitationsfelde wieder die Gleichungen (1) bzw. (1a); es ist aber dabei c nicht als eine Konstante, sondern als eine Funktion der Raumkoordinaten aufzufassen, die ein Maß für das Gravitationspotential darstellt. Aus (1a) folgen in bekannter Weise die Bewegungsgleichungen

$$(3) \quad \frac{d}{dt}\left\{ \frac{m\,\dot{x}}{\sqrt{c^2 - q^2}} \right\} = -\frac{m\,c\,\dfrac{\partial c}{\partial x}}{\sqrt{c^2 - q^2}}\,.$$

Man sieht, daß die Bewegungsgröße durch den nämlichen Ausdruck dargestellt wird wie oben. Überhaupt gelten für den im statischen Schwerefelde bewegten materiellen Punkt die Gleichungen (2). Die rechte Seite von (3) stellt die vom Gravitationsfelde auf den Massenpunkt ausgeübte Kraft $\Re_x$ dar. Für den Spezialfall der Ruhe ($q = 0$) ist

$$\Re_x = -m\,\frac{\partial c}{\partial x}\,.$$

Hieraus erkennt man, daß c die Rolle des Gravitationspotentials spielt Aus (2) folgt für einen langsam bewegten Punkt

$$(4) \quad \begin{aligned} J_x &= \frac{m\,\dot{x}}{c}\,, \\[2mm] E - mc &= \frac{\frac{1}{2}\,m\,q^2}{c}\,. \end{aligned}$$

Bei gegebener Geschwindigkeit sind also Impuls und kinetische Energie der Größe c umgekehrt proportional; anders ausgedrückt: Die träge Masse, so wie sie in Impuls und Energie eingeht, ist $\frac{m}{c}$, wobei m eine für den Massenpunkt charakteristische, vom Gravitationspotential unabhängige Konstante bedeutet. Es paßt dies zu Machs kühnem Gedanken, daß die Trägheit in einer Wechselwirkung des betrachteten Massenpunktes mit allen übrigen ihren Ursprung habe; denn häufen wir Massen in der Nähe des betrachteten Massenpunktes an, so verkleinern wir damit das Gravitationspotential c, erhöhen also die für die Trägheit maßgebende Größe $\frac{m}{c}$.

§ 2. Gleichungen für die Bewegung des materiellen Punktes im beliebigen Schwerefeld. Charakterisierung des letzteren.

Mit der Einführung einer räumlichen Veränderlichkeit der Größe c haben wir den Rahmen der gegenwärtig als „Relativitätstheorie" bezeichneten Theorie durchbrochen; denn es verhält sich nun der mit ds bezeichnete Ausdruck orthogonalenlinearen Transformationen der Koordinaten gegenüber nicht mehr als Invariante. Soll also — woran nicht zu zweifeln ist — das Relativitätsprinzip aufrecht erhalten werden, so müssen wir die Relativitätstheorie derart verallgemeinern, daß sie die im vorigen in ihren Elementen angedeutete Theorie des statischen Schwerefeldes als Spezialfall enthält.

Führen wir ein neues Raum-Zeitsystem $K'(x',y',z',t')$ ein durch irgend eine Substitution

$$x' = x'(x, y, z, t)$$
$$y' = y'(x, y, z, t)$$
$$z' = z'(x, y, z, t)$$
$$t' = t'(x, y, z, t).$$

und war das Schwerefeld im ursprünglichen System K ein statisches, so geht bei dieser Substitution die Gleichung (1) in eine Gleichung von der Form

$$\delta\left\{\int ds'\right\} = 0$$

über, wobei

$$ds'^2 = g_{11}\, dx'^2 + g_{22}\, dy'^2 + \ldots + 2\, g_{12}\, dx'\, dy' + \ldots$$

gesetzt ist, und die Größen $g_{\mu\nu}$ Funktionen von x', y', z', t' sind. Setzen wir x_1, x_2, x_3, x_4 statt x', y', z', t' und schreiben wir wieder ds statt ds', so erhalten die Bewegungsgleichungen des materiellen Punktes in bezug auf K' die Gestalt

$$(1'')\qquad \begin{cases} \delta\left\{\int ds\right\} = 0, \quad \text{wobei} \\ ds^2 = \sum_{\mu\nu} g_{\mu\nu}\, dx_\mu\, dx_\nu. \end{cases}$$

Wir gelangen so zu der Auffassung, daß im allgemeinen Falle das Gravitationsfeld durch zehn Raum-Zeit-Funktionen

$$
\begin{array}{cccc}
g_{11} & g_{12} & g_{13} & g_{14} \\
g_{21} & g_{22} & g_{23} & g_{24} \\
g_{31} & g_{32} & g_{33} & g_{34} \\
g_{41} & g_{42} & g_{43} & g_{44}
\end{array}
\qquad (g_{\mu\nu} = g_{\nu\mu})
$$

charakterisiert ist, welche sich im Falle der gewöhnlichen Relativitätstheorie auf

$$
\begin{array}{cccc}
-1 & 0 & 0 & 0 \\
0 & -1 & 0 & 0 \\
0 & 0 & -1 & 0 \\
0 & 0 & 0 & +c^2
\end{array}
$$

reduzieren, wobei c eine Konstante bedeutet. Dieselbe Art der Degeneration zeigt sich bei dem statischen Schwerefelde der vorhin betrachteten Art, nur daß bei diesem $g_{44} = c^2$ eine Funktion von x_1, x_2, x_3 ist.

Die Hamiltonsche Funktion H hat daher im allgemeinen Fall den Wert

$$(5) \quad H = -m \frac{ds}{dt} = -m\sqrt{g_{11}\dot{x}_1^2 + \cdots + 2g_{12}\dot{x}_1\dot{x}_2 + \cdots + 2g_{14}\dot{x}_1 + \cdots + g_{44}}.$$

Die zugehörigen Lagrangeschen Gleichungen

$$(6) \quad \frac{d}{dt}\left(\frac{\partial H}{\partial \dot{x}}\right) - \frac{\partial H}{\partial x} = 0$$

ergeben sofort den Ausdruck für den Impuls J des Punktes und für die vom Schwerefelde auf ihn ausgeübte Kraft $\mathfrak{K}$:

$$(7) \quad J_x = -m \frac{g_{11}\dot{x}_1 + g_{12}\dot{x}_2 + g_{13}\dot{x}_3 + g_{14}}{\frac{ds}{dt}} = -m \frac{g_{11}dx + g_{12}dx_2 + g_{13}dx_3 + g_{14}dx_4}{ds},$$

$$(8) \quad \mathfrak{K}_x = -\tfrac{1}{2}m \frac{\sum_{\mu\nu}\frac{\partial g_{\mu\nu}}{\partial x_1}dx_\mu dx_\nu}{ds \cdot dt} = -\tfrac{1}{2}m \cdot \sum_{\mu\nu}\frac{\partial g_{\mu\nu}}{\partial x_1} \cdot \frac{dx_\mu}{ds} \cdot \frac{dx_\nu}{dt}.$$

Ferner ergibt sich für die Energie E des Punktes

$$(9) \quad -E = -\left(\dot{x}\frac{\partial H}{\partial \dot{x}} + \cdots\right) + H = -m\left(g_{41}\frac{dx_1}{ds} + g_{42}\frac{dx_2}{ds} + g_{43}\frac{dx_3}{ds} + g_{44}\frac{dx_4}{ds}\right).$$

Im Falle der gewöhnlichen Relativitätstheorie sind nur lineare orthogonale Substitutionen zulässig. Es wird sich zeigen, daß wir für die Einwirkung des Schwerefeldes auf die materiellen Vorgänge Gleichungen aufzustellen vermögen, die beliebigen Substitutionen gegenüber sich kovariant verhalten.

Zunächst können wir aus der Bedeutung, welche ds im Bewegungs-
gesetz des materiellen Punktes spielt, den Schluß ziehen, daß ds eine
absolute Invariante (Skalar) sein muß; hieraus ergibt sich, daß die
Größen $g_{\mu\nu}$ einen kovarianten Tensor zweiten Ranges bilden[1]), den
wir als den kovarianten Fundamentaltensor bezeichnen. Dieser bestimmt
das Schwerefeld. Es ergibt sich ferner aus (7) und (9), daß Impuls und
Energie des materiellen Punktes zusammen einen kovarianten Tensor
ersten Ranges, d. h. einen kovarianten Vektor bilden.[2])

§ 3. Bedeutung des Fundamentaltensors der $g_{\mu\nu}$ für die Messung von Raum und Zeit.

Aus dem Früheren kann man schon entnehmen, daß zwischen den
Raum-Zeit-Koordinaten x_1, x_2, x_3, x_4 und den mittelst Maßstäben und
Uhren zu erhaltenden Meßergebnissen keine so einfachen Beziehungen
bestehen können, wie in der alten Relativitätstheorie. Es ergab sich
dies bezüglich der Zeit schon beim statischen Schwerefelde.[3]) Es erhebt
sich deshalb die Frage nach der physikalischen Bedeutung (prinzipiellen
Meßbarkeit) der Koordinaten x_1, x_2, x_3, x_4.

Hierzu bemerken wir, daß ds als invariantes Maß für den Abstand
zweier unendlich benachbarter Raumzeitpunkte aufzufassen ist. Es muß
daher ds auch eine vom gewählten Bezugssystem unabhängige physikalische Bedeutung zukommen. Wir nehmen an, ds sei der „natürlich
gemessene" Abstand beider Raumzeitpunkte und wollen darunter fol-
gendes verstehen.

Die unmittelbare Nachbarschaft des Punktes (x_1, x_2, x_3, x_4) wird bezüglich des Koordinatensystems durch die infinitesimalen Variabeln dx_1,
dx_2, dx_3, dx_4 bestimmt. Wir denken uns statt dieser durch eine lineare
Transformation neue Variable $d\xi_1$, $d\xi_2$, $d\xi_3$, $d\xi_4$ eingeführt, derart, daß

$$ds^2 = d\xi_1^2 + d\xi_2^2 + d\xi_3^2 - d\xi_4^2$$

wird. Bei dieser Transformation sind die $g_{\mu\nu}$ als Konstanten zu betrachten; der reelle Kegel $ds^2 = 0$ erscheint auf seine Hauptachsen bezogen. In diesem elementaren $d\xi$-System gilt dann die gewöhnliche
Relativitätstheorie, und es sei in diesem System die physikalische Bedeutung von Längen und Zeiten dieselbe wie in der gewöhnlichen Relativitätstheorie. d. h. ds^2 ist das Quadrat des vierdimensionalen Abstandes beider unendlich benachbarter Raumzeitpunkte, gemessen mittelst
eines im $d\xi$-System nicht beschleunigten starren Körpers und mittelst
relativ zu diesem ruhend angeordneter Einheitsmaßstäbe und Uhren.

---

1) Vgl. II. Teil, § 1.   2) Vgl. II. Teil, § 1.
3) Vgl. z. B. A. Einstein, Ann. d. Phys. 4. 35. S. 903 ff.

Man sieht hieraus, daß bei gegebenen dx_1, dx_2, dx_3, dx_4 der zu diesen Differentialen gehörige natürliche Abstand nur dann ermittelt werden kann, wenn die das Gravitationsfeld bestimmenden Größen $g_{\mu\nu}$ bekannt sind. Man kann dies auch so ausdrücken: Das Gravitationsfeld beeinflußt die Meßkörper und Uhren in bestimmter Weise.

Aus der Fundamentalgleichung

$$ds^2 = \sum_{\mu\nu} g_{\mu\nu} dx_\mu dx_\nu$$

sieht man, daß es zur Festlegung der physikalischen Dimension der Größen $g_{\mu\nu}$ und x_ν noch einer Festsetzung bedarf. Der Größe ds kommt die Dimension einer Länge zu. Wir wollen die x_ν ebenfalls als Längen ansehen (auch x_4), den Größen $g_{\mu\nu}$ also keine physikalische Dimension zuschreiben.

§ 4. Bewegung kontinuierlich verteilter inkohärenter Massen im beliebigen Schwerefeld.

Zur Ableitung des Bewegungsgesetzes kontinuierlich verteilter inkohärenter Massen berechnen wir Impuls und ponderomotorische Kraft pro Volumeneinheit und wenden hierauf den Impulssatz an.

Dazu haben wir zunächst das dreidimensionale Volumen V unseres Massenpunktes zu berechnen. Wir betrachten ein unendlich kleines (vierdimensionales) Stück des Raumzeitfadens unseres materiellen Punktes. Sein Volumen ist

$$\int\int\int\int dx_1 dx_2 dx_3 dx_4 = V dt.$$

Führen wir statt der dx die natürlichen Differentiale $d\xi$ ein, wobei der Meßkörper als gegen den materiellen Punkt ruhend angenommen wird, so haben wir

$$\int\int\int d\xi_1 d\xi_2 d\xi_3 = V_0$$

zu setzen, d. h. gleich dem „Ruhvolumen" des materiellen Punktes. Ferner haben wir

$$\int d\xi_4 = ds,$$

wo ds dieselbe Bedeutung hat wie oben.

Sind die dx mit den $d\xi$ verbunden durch die Substitution

$$dx_\mu = \sum_\sigma \alpha_{\mu\sigma} d\xi_\sigma,$$

so hat man

$$\int\int\int\int dx_1 dx_2 dx_3 dx_4 = \int\int\int\int \frac{\partial(dx_1, dx_2, dx_3, dx_4)}{\partial(d\xi_1, d\xi_2, d\xi_3, d\xi_4)} \cdot d\xi_1 d\xi_2 d\xi_3 d\xi_4$$

oder

$$V dt = V_0 ds \cdot |\alpha_{\rho\sigma}|.$$

Da aber

$$ds^2 = \sum_{\mu\nu} g_{\mu\nu} dx_\mu dx_\nu = \sum_{\mu\nu\varrho\sigma} g_{\mu\nu}\alpha_{\mu\varrho}\alpha_{\nu\sigma} d\xi_\varrho d\xi_\sigma = d\xi_1^2 + d\xi_2^2 + d\xi_3^2 - d\xi_4^2$$

ist, so besteht zwischen der Determinante

$$g = |g_{\mu\nu}|,$$

d. h. der Diskriminante der quadratischen Differentialform ds^2 und der Substitutionsdeterminante $|\alpha_{\varrho\sigma}|$ die Beziehung

$$g \cdot (|\alpha_{\varrho\sigma}|)^2 = -1,$$

$$|\alpha_{\varrho\sigma}| = \frac{1}{\sqrt{-g}}.$$

Man erhält also für V die Beziehung

$$V dt = V_0 ds \cdot \frac{1}{\sqrt{-g}}.$$

Hieraus ergibt sich mit Hilfe von (7), (8) und (9), wenn man $\frac{m}{V_0}$ durch ϱ_0 ersetzt

$$\frac{J_x}{V} = -\varrho_0\sqrt{-g} \cdot \sum_\nu g_{1\nu} \frac{dx_\nu}{ds} \cdot \frac{dx_4}{ds},$$

$$-\frac{E}{V} = -\varrho_0\sqrt{-g} \cdot \sum_\nu g_{4\nu} \frac{dx_\nu}{ds} \cdot \frac{dx_4}{ds},$$

$$\frac{\Re_x}{V} = -\frac{1}{2}\varrho_0\sqrt{-g} \cdot \sum_{\mu\nu} \frac{\partial g_{\mu\nu}}{\partial x_1} \cdot \frac{dx_\mu}{ds} \cdot \frac{dx_\nu}{ds}.$$

Wir bemerken, daß

$$\Theta_{\mu\nu} = \varrho_0 \frac{dx_\mu}{ds} \cdot \frac{dx_\nu}{ds}$$

ein kontravarianter Tensor zweiten Ranges bezüglich beliebiger Substitutionen ist. Man vermutet aus dem Vorhergehenden, daß der Impuls-Energiesatz die Form haben wird:

$$(10) \quad \sum_{\mu\nu} \frac{\partial}{\partial x_\nu}(\sqrt{-g} \cdot g_{\sigma\mu}\Theta_{\mu\nu}) - \frac{1}{2}\sum_{\mu\nu}\sqrt{-g} \cdot \frac{\partial g_{\mu\nu}}{\partial x_\sigma}\Theta_{\mu\nu} = 0. \qquad (\sigma = 1,2,3,4)$$

Die ersten drei dieser Gleichungen ($\sigma = 1, 2, 3$) drücken den Impulssatz, die letzte ($\sigma = 4$) den Energiesatz aus. Es erweist sich in der Tat, daß diese Gleichungen beliebigen Substitutionen gegenüber kovariant sind.[1] Ferner lassen sich die Bewegungsgleichungen des materiellen Punktes, von denen wir ausgegangen sind, aus diesen Gleichungen durch Integration über den Stromfaden wieder ableiten.

---

[1] Vgl. II. Teil, § 4, Nr. 1.

Den Tensor $\Theta_{\mu\nu}$ nennen wir den (kontravarianten) Spannungs-Energietensor der materiellen Strömung. Der Gleichung (10) schreiben wir einen Gültigkeitsbereich zu, der über den speziellen Fall der Strömung inkohärenter Massen weit hinausgeht. Die Gleichung stellt allgemein die Energiebilanz zwischen dem Gravitationsfelde und einem beliebigen materiellen Vorgang dar; nur ist für $\Theta_{\mu\nu}$ der dem jeweilen betrachteten materiellen System entsprechende Spannungs-Energietensor einzusetzen. Die erste Summe in der Gleichung enthält die örtlichen Ableitungen der Spannungen bzw. Energiestromdichte und die zeitlichen Ableitungen der Impuls- bzw. Energiedichte; die zweite Summe ist ein Ausdruck für die Wirkungen, welche vom Schwerefelde auf den materiellen Vorgang übertragen werden.

§ 5. Die Differentialgleichungen des Gravitationsfeldes.

Nachdem wir die Impuls-Energiegleichung für die materiellen Vorgänge (mechanische, elektrische und andere Vorgänge) mit bezug auf das Gravitationsfeld aufgestellt haben, bleibt uns noch folgende Aufgabe. Es sei der Tensor $\Theta_{\mu\nu}$ für den materiellen Vorgang gegeben. Welches sind die Differentialgleichungen, welche die Größen g_{ik}, d. h. das Schwerefeld zu bestimmen gestatten? Wir suchen mit anderen Worten die Verallgemeinerung der Poissonschen Gleichung

$$\Delta\varphi = 4\pi k\varrho.$$

Zur Lösung dieser Aufgabe haben wir keine so vollkommen zwangläufige Methode gefunden, wie für die Lösung des vorhin behandelten Problems. Es war nötig, einige Annahmen einzuführen, deren Richtigkeit zwar plausibel erscheint, aber doch nicht evident ist.

Die gesuchte Verallgemeinerung wird wohl von der Form sein

(11) $$\varkappa \cdot \Theta_{\mu\nu} = \Gamma_{\mu\nu},$$

wo $\varkappa$ eine Konstante, $\Gamma_{\mu\nu}$ ein kontravarianter Tensor zweiten Ranges ist, der durch Differentialoperationen aus dem Fundamentaltensor $g_{\mu\nu}$ hervorgeht. Dem Newton-Poissonschen Gesetz entsprechend wird man geneigt sein zu fordern, daß diese Gleichungen (11) zweiter Ordnung sein sollen. Es muß aber hervorgehoben werden, daß es sich als unmöglich erweist, unter dieser Voraussetzung einen Differentialausdruck $\Gamma_{\mu\nu}$ zu finden, der eine Verallgemeinerung von $\Delta\varphi$ ist, und sich beliebigen Transformationen gegenüber als Tensor erweist.[1] A priori kann allerdings nicht in Abrede gestellt werden, daß die endgültigen, genauen Gleichungen der Gravitation von höherer als zweiter Ordnung sein könnten. Es besteht daher immer noch die Möglichkeit, daß die

---

1) Vgl. II. Teil, § 4, Nr. 2.

vollkommen exakten Differentialgleichungen der Gravitation beliebigen
Substitutionen gegenüber kovariant sein könnten. Der Versuch einer
Diskussion derartiger Möglichkeiten wäre aber bei dem gegenwärtigen
Stande unserer Kenntnis der physikalischen Eigenschaften des Gravitationsfeldes verfrüht. Deshalb ist für uns die Beschränkung auf die
zweite Ordnung geboten und wir müssen daher darauf verzichten, Gravitationsgleichungen aufzustellen, die sich beliebigen Transformationen
gegenüber als kovariant erweisen. Es ist übrigens hervorzuheben, daß
wir keinerlei Anhaltspunkte für eine allgemeine Kovarianz der Gravitationsgleichungen haben.[1])

Der Laplacesche Skalar $\Delta\varphi$ ergibt sich aus dem Skalar φ, indem
man von diesem die Erweiterung (den Gradienten), und dann von
diesem den inneren Operator (die Divergenz) bildet. Beide Operationen
kann man derart verallgemeinern, daß sie an jedem Tensor von beliebig
hohem Rang ausgeführt werden können, und zwar unter Zulassung
beliebiger Substitutionen der Grundvariabeln.[2]) Aber es degenerieren
diese Operationen, wenn sie an dem Fundamentaltensor $g_{\mu\nu}$ ausgeführt
werden.[3]) Es scheint daraus hervorzugehen, daß die gesuchten Gleichungen nur bezüglich einer gewissen Gruppe von Transformationen
kovariant sein werden, welche Gruppe uns aber vorläufig unbekannt ist.

Bei dieser Sachlage erscheint es mit Rücksicht auf die alte Relativitätstheorie natürlich, anzunehmen, daß in der gesuchten Transformationsgruppe die linearen Transformationen enthalten
seien. Wir fordern also, daß $\Gamma_{\mu\nu}$ ein Tensor bezüglich beliebiger
linearer Transformationen sein soll.

Man beweist nun leicht (durch Ausführung der Transformation)
die folgenden Sätze:

1. Ist $\Theta_{\alpha\beta\ldots\lambda}$ ein kontravarianter Tensor vom Range n bezüglich
linearer Transformationen, so ist

$$\sum_\mu \gamma_{\mu\nu}\frac{\partial\Theta_{\alpha\beta\ldots\lambda}}{\partial x_\mu}$$

ein kontravarianter Tensor vom Range $n+1$ bezüglich linearer Transformationen (Erweiterung).[4])

2. Ist $\Theta_{\alpha\beta\ldots\lambda}$ ein kontravarianter Tensor vom Range n bezüglich
linearer Transformationen, so ist

$$\sum_\lambda \frac{\partial\Theta_{\alpha\beta\ldots\lambda}}{\partial x_\lambda}$$

---

1) Vgl. hierzu noch die am Anfange des § 6 gegebenen Überlegungen.
2) II. Teil, § 2.   3) Vgl. die Anm. auf S. 28 im II. Teil, § 2.
4) $\gamma_{\mu\nu}$ ist der zu $g_{\mu\nu}$ reziproke kontravariante Tensor (II. Teil, § 1).

ein kontravarianter Tensor vom Range $n-1$ bezüglich linearer Transformationen (Divergenz).

Führt man an einem Tensor der Reihe nach diese beiden Operationen aus, so erhält man einen Tensor, der wiederum vom gleichen Range ist, wie der ursprüngliche (Operation Δ, an einem Tensor vorgenommen). Für den Fundamental-Tensor $\gamma_{\mu\nu}$ erhält man

(a)
$$\sum_{\alpha\beta}\frac{\partial}{\partial x_\alpha}\left(\gamma_{\alpha\beta}\frac{\partial\gamma_{\mu\nu}}{\partial x_\beta}\right).$$

Daß dieser Operator mit dem Laplaceschen Operator verwandt ist, erkennt man ferner durch folgende Betrachtung. In der Relativitätstheorie (Fehlen des Gravitationsfeldes), wäre zu setzen

$$g_{11}=g_{22}=g_{33}=-1,\quad g_{44}=c^2,\quad g_{\mu\nu}=0,\ \text{für}\ \mu\neq\nu;$$

also

$$\gamma_{11}=\gamma_{22}=\gamma_{33}=-1,\quad \gamma_{44}=\frac{1}{c^2},\ \gamma_{\mu\nu}=0,\ \text{für}\ \mu\neq\nu.$$

Ist ein Gravitationsfeld vorhanden, welches genügend schwach ist, d. h. unterscheiden sich die $g_{\mu\nu}$ und $\gamma_{\mu\nu}$ von den soeben angegebenen Werten nur unendlich wenig, so erhält man an Stelle des Ausdruckes (a) unter Vernachlässigung der Glieder vom zweiten Grade

$$-\left(\frac{\partial^2\gamma_{\mu\nu}}{\partial x_1^2}+\frac{\partial^2\gamma_{\mu\nu}}{\partial x_2^2}+\frac{\partial^2\gamma_{\mu\nu}}{\partial x_3^2}-\frac{1}{c^2}\frac{\partial^2\gamma_{\mu\nu}}{\partial x_4^2}\right).$$

Ist das Feld ein statisches und nur g_{44} variabel, so kommen wir also auf den Fall der Newtonschen Gravitationstheorie, falls wir den gebildeten Ausdruck bis auf eine Konstante für die Größe $\Gamma_{\mu\nu}$ setzen.

Man könnte demnach denken, es müsse der Ausdruck (a) bis auf einen konstanten Faktor bereits die gesuchte Verallgemeinerung von $\Delta\varphi$ sein. Dies wäre aber ein Irrtum; denn es könnten neben jenem Ausdruck noch solche Terme in einer derartigen Verallgemeinerung auftreten, die selbst Tensoren sind und bei Durchführung der eben angeführten Vernachlässigungen verschwinden. Es tritt dies immer dann ein, wenn zwei erste Ableitungen der $g_{\mu\nu}$ bzw. $\gamma_{\mu\nu}$ miteinander multipliziert erscheinen. So ist z. B.

$$\sum_{\alpha\beta}\frac{\partial g_{\alpha\beta}}{\partial x_\mu}\cdot\frac{\partial\gamma_{\alpha\beta}}{\partial x_\nu}$$

ein kovarianter Tensor zweiten Ranges (gegenüber linearen Transformationen); derselbe wird unendlich klein zweiter Ordnung, wenn die Größen $g_{\alpha\beta}$ und $\gamma_{\alpha\beta}$ von Konstanten nur um Unendlich-Kleine erster Ordnung abweichen. Wir müssen daher zulassen, daß in $\Gamma_{\mu\nu}$ neben (a) noch andere Terme auftreten, die vorläufig nur die Bedingung erfüllen

müssen, daß sie zusammen linearen Transformationen gegenüber Tensor-
charakter besitzen müssen.

Zur Auffindung dieser Terme dient uns der Impulsenergiesatz.
Damit die benutzte Methode klar hervortrete, will ich sie zunächst an
einem allgemein bekannten Beispiel anwenden.

In der Elektrostatik ist $-\dfrac{\partial \varphi}{\partial x_\nu}\varrho$ die ν^{te} Komponente des pro
Volumeneinheit auf die Materie übertragenen Impulses, falls φ das elek-
trostatische Potential, ϱ die elektrische Dichte bedeutet. Es ist eine
Differentialgleichung für φ gesucht, derart, daß der Impulssatz stets
erfüllt ist. Es ist wohlbekannt, daß die Gleichung

$$\sum_\nu \frac{\partial^2 \varphi}{\partial x_\nu^2} = \varrho$$

die Aufgabe löst. Daß der Impulssatz erfüllt ist, geht hervor aus der
Identität

$$\sum_\mu \frac{\partial}{\partial x_\mu}\left(\frac{\partial \varphi}{\partial x_\nu}\frac{\partial \varphi}{\partial x_\mu}\right) - \frac{\partial}{\partial x_\nu}\left(\frac{1}{2}\sum_\mu \left(\frac{\partial \varphi}{\partial x_\mu}\right)^2\right) = \frac{\partial \varphi}{\partial x_\nu}\sum_\mu \frac{\partial^2 \varphi}{\partial x_\mu^2}\left(= -\frac{\partial \varphi}{\partial x_\nu}\cdot\varrho\right).$$

Wenn also der Impulssatz erfüllt ist, muß für jedes ν eine identische Gleichung von folgendem Bau existieren: Auf der rechten Seite
steht $-\dfrac{\partial \varphi}{\partial x_\nu}$ multipliziert mit der linken Seite der Differentialgleichung, auf
der linken Seite der Identität steht eine Summe von Differentialquotienten.

Wäre die Differentialgleichung für φ noch nicht bekannt, so ließe
sich das Problem von deren Auffindung auf dasjenige der Auffindung jener
identischen Gleichung zurückführen. Es ist nun für uns die Erkenntnis
wesentlich, daß jene Identität sich ableiten läßt, wenn einer der in
ihr auftretenden Terme bekannt ist. Man hat nichts weiteres zu
tun, als die Regel von der Differentiation eines Produktes in den Formen

$$\frac{\partial}{\partial x_\nu}(uv) = \frac{\partial u}{\partial x_\nu}v + \frac{\partial v}{\partial x_\nu}u$$

und

$$u\frac{\partial v}{\partial x_\nu} = \frac{\partial}{\partial x_\nu}(uv) - \frac{\partial u}{\partial x_\nu}v$$

wiederholt anzuwenden und schließlich die Glieder, welche Differentialquotienten sind, auf die linke Seite, die übrigen auf die rechte Seite
zu stellen. Geht man z. B. von dem ersten Glied der obigen Identität
aus, so erhält man der Reihe nach

$$\sum_\mu \frac{\partial}{\partial x_\mu}\left(\frac{\partial \varphi}{\partial x_\nu}\frac{\partial \varphi}{\partial x_\mu}\right) = \sum_\mu \frac{\partial \varphi}{\partial x_\nu}\cdot\frac{\partial^2 \varphi}{\partial x_\mu^2} + \sum_\mu \frac{\partial \varphi}{\partial x_\mu}\cdot\frac{\partial^2 \varphi}{\partial x_\nu \partial x_\mu}$$

$$= \frac{\partial \varphi}{\partial x_\nu}\cdot\sum_\mu \frac{\partial^2 \varphi}{\partial x_\mu^2} + \frac{\partial}{\partial x_\nu}\left\{\frac{1}{2}\sum_\mu \left(\frac{\partial \varphi}{\partial x_\mu}\right)^2\right\},$$

woraus durch Anordnen die obige Identität hervorgeht.

Wir wenden uns nun unserem Problem wieder zu. Aus Gleichung (10) geht hervor, daß

$$\frac{1}{2}\sum_{\mu\nu}\sqrt{-g}\cdot\frac{\partial g_{\mu\nu}}{\partial x_a}\,\Theta_{\mu\nu}. \qquad (a=1,2,3,4)$$

der pro Volumeneinheit auf die Materie vom Gravitationsfeld übertragene Impuls (bzw. Energie) ist. Damit der Energie-Impulssatz erfüllt sei, müssen die Differentialausdrücke $\Gamma_{\mu\nu}$ der Fundamentalgrößen $g_{\mu\nu}$, welche in die Gravitationsgleichungen

$$\varkappa\cdot\Theta_{\mu\nu}=\Gamma_{\mu\nu}$$

eingehen, so gewählt werden, daß

$$\frac{1}{2\varkappa}\sum_{\mu\nu}\sqrt{-g}\cdot\frac{\partial g_{\mu\nu}}{\partial x_a}\,\Gamma_{\mu\nu}$$

sich derart umformen läßt, daß er als Summe von Differentialquotienten erscheint. Es ist andererseits bekannt, daß in dem für $\Gamma_{\mu\nu}$ zu suchenden Ausdruck der Term (a) erscheint. Die gesuchte identische Gleichung ist also von folgender Gestalt:

Summe von Differentialquotienten

$$=\frac{1}{2}\sum_{\mu\nu}\sqrt{-g}\cdot\frac{\partial g_{\mu\nu}}{\partial x_a}\left\{\sum_{a,\beta}\frac{\partial}{\partial x_\beta}\left(\gamma_{a\beta}\frac{\partial\gamma_{\mu\nu}}{\partial x_\beta}\right)\right.$$

+ weitere Glieder, die bei Bildung der ersten Annäherung wegfallen.}

Hierdurch ist die gesuchte Identität eindeutig bestimmt; bildet man sie nach dem angedeuteten Verfahren[1]), so erhält man:

$$(12)\begin{cases}\displaystyle\sum_{a\beta\tau\varrho}\frac{\partial}{\partial x_a}\left(\sqrt{-g}\cdot\gamma_{a\beta}\frac{\partial\gamma_{\tau\varrho}}{\partial x_\beta}\frac{\partial g_{\tau\varrho}}{\partial x_a}\right)-\frac{1}{2}\cdot\sum_{a\beta\tau\varrho}\frac{\partial}{\partial x_a}\left(\sqrt{-g}\cdot\gamma_{a\beta}\frac{\partial\gamma_{\tau\varrho}}{\partial x_a}\frac{\partial g_{\tau\varrho}}{\partial x_\beta}\right)\\[2mm]\displaystyle=\sum_{\mu\nu}\sqrt{-g}\frac{\partial g_{\mu\nu}}{\partial x_a}\left\{\sum_{a\beta}\frac{1}{\sqrt{-g}}\frac{\partial}{\partial x_a}\left(\gamma_{a\beta}\sqrt{-g}\cdot\frac{\partial\gamma_{\mu\nu}}{\partial x_\beta}\right)-\sum_{a\beta\tau\varrho}\gamma_{a\beta}g_{\tau\varrho}\frac{\partial\gamma_{\mu\tau}}{\partial x_a}\frac{\partial\gamma_{\nu\varrho}}{\partial x_\beta}\right.\\[2mm]\displaystyle\qquad\left.+\frac{1}{2}\sum_{a\beta\tau\varrho}\gamma_{a\mu}\gamma_{\beta\nu}\frac{\partial\gamma_{\tau\varrho}}{\partial x_a}\frac{\partial\gamma_{\tau\varrho}}{\partial x_\beta}-\frac{1}{2}\sum_{a\beta\tau\varrho}\gamma_{\mu\nu}\gamma_{a\beta}\frac{\partial g_{\tau\varrho}}{\partial x_a}\frac{\partial\gamma_{\tau\varrho}}{\partial x_\beta}\right\}.\end{cases}$$

Der in der geschweiften Klammer der rechten Seite stehende Ausdruck $\Gamma_{\mu\nu}$ ist demnach der von uns gesuchte Tensor, der in die Gravitationsgleichungen

$$\varkappa\Theta_{\mu\nu}=\Gamma_{\mu\nu}$$

eintritt. Um diese Gleichungen besser überblicken zu können, führen wir folgende Abkürzungen ein:

$$(13)\qquad -2\varkappa\cdot\mathfrak{t}_{\mu\nu}=\sum_{a\beta\tau\varrho}\left(\gamma_{a\mu}\gamma_{\beta\nu}\frac{\partial g_{\tau\varrho}}{\partial x_a}\cdot\frac{\partial\gamma_{\tau\varrho}}{\partial x_\beta}-\frac{1}{2}\gamma_{\mu\nu}\gamma_{a\beta}\frac{\partial g_{\tau\varrho}}{\partial x_a}\frac{\partial\gamma_{\tau\varrho}}{\partial x_\beta}\right).$$

1) Vgl. II. Teil. § 4, Nr. 3.

$\vartheta_{\mu\nu}$ sei als „kontravarianter Spannungs-Energietensor des Gravitationsfeldes" bezeichnet. Den zu ihm reziproken kovarianten Tensor bezeichnen wir mit $t_{\mu\nu}$; es ist also

$$(14) \qquad -2\varkappa \cdot t_{\mu\nu} = \sum_{\alpha\beta\tau\varrho} \left(\frac{\partial g_{\tau\varrho}}{\partial x_\mu} \frac{\partial \gamma_{\tau\varrho}}{\partial x_\nu} - \tfrac{1}{2} g_{\mu\nu} \gamma_{\alpha\beta} \frac{\partial g_{\tau\varrho}}{\partial x_\alpha} \frac{\partial \gamma_{\tau\varrho}}{\partial x_\beta} \right).$$

Ebenfalls zur Abkürzung führen wir folgende Bezeichnungen ein für Differentialoperationen, ausgeführt an den Fundamentaltensoren γ bzw. g:

$$(15) \quad \varDelta_{\mu\nu}(\gamma) = \sum_{\alpha\beta} \frac{1}{\sqrt{-g}} \cdot \frac{\partial}{\partial x_\alpha} \left(\gamma_{\alpha\beta} \sqrt{-g} \cdot \frac{\partial \gamma_{\mu\nu}}{\partial x_\beta} \right) - \sum_{\alpha\beta\tau\varrho} \gamma_{\alpha\beta} g_{\tau\varrho} \frac{\partial \gamma_{\mu\tau}}{\partial x_\alpha} \frac{\partial \gamma_{\nu\varrho}}{\partial x_\beta},$$

bzw.

$$(16) \quad D_{\mu\nu}(g) = \sum_{\alpha\beta} \frac{1}{\sqrt{-g}} \cdot \frac{\partial}{\partial x_\alpha} \left(\gamma_{\alpha\beta} \sqrt{-g} \cdot \frac{\partial g_{\mu\nu}}{\partial x_\beta} \right) - \sum_{\alpha\beta\tau\varrho} \gamma_{\alpha\beta} \gamma_{\tau\varrho} \frac{\partial g_{\mu\tau}}{\partial x_\alpha} \frac{\partial g_{\nu\varrho}}{\partial x_\beta}.$$

Jeder dieser Operatoren liefert wieder einen Tensor der gleichen Art (bezügl. linearer Transformationen).

Bei Verwendung dieser Abkürzungen nimmt die Identität (12) die Form an:

$$(12\,\text{a}) \sum_{\mu\nu} \frac{\partial}{\partial x_\nu} \left\{ \sqrt{-g} \cdot g_{\sigma\mu} \cdot \varkappa \vartheta_{\mu\nu} \right\} = \tfrac{1}{2} \sum_{\mu\nu} \sqrt{-g} \cdot \frac{\partial g_{\mu\nu}}{\partial x_\sigma} \left\{ -\varDelta_{\mu\nu}(\gamma) + \varkappa \vartheta_{\mu\nu} \right\},$$

oder auch

$$(12\,\text{b}) \sum_{\mu\nu} \frac{\partial}{\partial x_\nu} \left\{ \sqrt{-g} \cdot \gamma_{\mu\nu} \cdot \varkappa t_{\mu\sigma} \right\} = \tfrac{1}{2} \sum_{\mu\nu} \sqrt{-g} \cdot \frac{\partial \gamma_{\mu\nu}}{\partial x_\sigma} \left\{ -D_{\mu\nu}(g) - \varkappa \cdot t_{\mu\nu} \right\}.$$

Schreiben wir die Erhaltungsgleichung (10) der Materie und die Erhaltungsgleichung (12a) für das Gravitationsfeld in der Form

$$(10)' \qquad \sum_{\mu\nu} \frac{\partial}{\partial x_\nu} \left(\sqrt{-g} \cdot g_{\sigma\mu} \cdot \Theta_{\mu\nu} \right) - \tfrac{1}{2} \sum_{\mu\nu} \sqrt{-g} \cdot \frac{\partial g_{\mu\nu}}{\partial x_\sigma} \cdot \Theta_{\mu\nu} = 0$$

$$(12\,\text{c}) \qquad \begin{aligned} &\sum_{\mu\nu} \frac{\partial}{\partial x_\nu} \left(\sqrt{-g} \cdot g_{\sigma\mu} \cdot \vartheta_{\mu\nu} \right) - \tfrac{1}{2} \sum_{\mu\nu} \sqrt{-g} \cdot \frac{\partial g_{\mu\nu}}{\partial x_\sigma} \cdot \vartheta_{\mu\nu} \\ &\qquad = -\frac{1}{2\varkappa} \cdot \sum_{\mu\nu} \sqrt{-g} \cdot \frac{\partial g_{\mu\nu}}{\partial x_\sigma} \cdot \varDelta_{\mu\nu}(\gamma), \end{aligned}$$

so erkennt man, daß der Spannungs-Energie-Tensor $\vartheta_{\mu\nu}$ des Gravitationsfeldes in den Erhaltungssatz für das Gravitationsfeld genau ebenso eintritt, wie der Tensor $\Theta_{\mu\nu}$ des materiellen Vorganges in den Erhaltungssatz für diesen Vorgang, ein bemerkenswerter Umstand bei der Verschiedenheit der Ableitungen beider Sätze.

Aus der Gleichung (12a) folgt als Ausdruck für den Differentialtensor, der in die Gravitationsgleichungen eingeht

(17) $$\Gamma_{\mu\nu} = \varDelta_{\mu\nu}(\gamma) - \varkappa \cdot \vartheta_{\mu\nu}.$$

Die Gravitationsgleichungen (11) lauten also

(18) $$\varDelta_{\mu\nu}(\gamma) = \varkappa(\Theta_{\mu\nu} + \vartheta_{\mu\nu}).$$

Diese Gleichungen erfüllen eine Forderung, die unseres Erachtens an eine Relativitätstheorie der Gravitation notwendig gestellt werden muß; sie zeigen nämlich, daß der Tensor $\vartheta_{\mu\nu}$ des Gravitationsfeldes in gleicher Weise felderregend auftritt, wie der Tensor $\Theta_{\mu\nu}$ der materiellen Vorgänge. Eine Ausnahmestellung der Gravitationsenergie gegenüber allen anderen Energiearten würde ja zu unhaltbaren Konsequenzen führen.

Durch Addition der Gleichungen (10) und (12a) findet man mit Rücksicht auf die Gleichung (18)

(19) $$\sum_{\mu\nu} \frac{\partial}{\partial x_\nu}\left\{\sqrt{-g}\cdot g_{\sigma\mu}(\Theta_{\mu\nu} + \vartheta_{\mu\nu})\right\} = 0. \qquad (\sigma = 1, 2, 3, 4)$$

Hieraus ersieht man, daß für Materie und Gravitationsfeld zusammen die Erhaltungssätze gelten.

Bei der bisher gegebenen Darstellung haben wir die kontravarianten Tensoren bevorzugt, weil sich der kontravariante Spannungsenergietensor der Strömung inkohärenter Massen in besonders einfacher Weise ausdrücken läßt. Indessen können wir die gewonnenen Fundamentalbeziehungen ebenso einfach unter Benutzung kovarianter Tensoren ausdrücken. Statt $\Theta_{\mu\nu}$ haben wir dann $T_{\mu\nu} = \sum_{\alpha\beta} g_{\mu\alpha} g_{\nu\beta}\Theta_{\alpha\beta}$ als SpannungsEnergietensor des materiellen Vorganges zugrunde zu legen. Statt Gleichung (10) erhalten wir durch gliedweise Umformung

(20) $$\sum_{\mu\nu} \frac{\partial}{\partial x_\nu}\left(\sqrt{-g}\cdot\gamma_{\mu\nu} T_{\mu\sigma}\right) + \frac{1}{2}\sum_{\mu\nu}\sqrt{-g}\cdot\frac{\partial\gamma_{\mu\nu}}{\partial x_\sigma}\cdot T_{\mu\nu} = 0.$$

Aus dieser Gleichung und (16) folgt, daß die Gleichungen des Gravitationsfeldes auch in der Form

(21) $$- D_{\mu\nu}(g) = \varkappa(t_{\mu\nu} + T_{\mu\nu})$$

geschrieben werden können, welche Gleichungen auch direkt aus (18) abgeleitet werden können. Analog (19) besteht die Beziehung

(22) $$\sum_\nu \frac{\partial}{\partial x_\nu}\left\{\sqrt{-g}\cdot\gamma_{\sigma\mu}(T_{\mu\nu} + t_{\mu\nu})\right\} = 0.$$

§ 6. Einfluß des Gravitationsfeldes auf physikalische Vorgänge, speziell auf die elektromagnetischen Vorgänge.

Weil bei jeglichem physikalischen Vorgang Impuls und Energie eine Rolle spielen, diese letzteren aber ihrerseits das Gravitationsfeld

bestimmen und von ihm beeinflußt werden, müssen die das Schwerefeld
bestimmenden Größen $g_{\mu\nu}$ in allen physikalischen Gleichungssystemen
auftreten. So haben wir gesehen, daß die Bewegung des materiellen
Punktes durch die Gleichung

$$\delta\left\{\int ds\right\} = 0$$

bestimmt ist, wobei

$$ds^2 = \sum_{\mu\nu} g_{\mu\nu}\, dx_\mu dx_\nu.$$

ds ist eine Invariante beliebigen Substitutionen gegenüber. Die gesuchten Gleichungen, welche den Ablauf irgend eines physikalischen
Vorganges bestimmen, müssen nun so gebaut sein, daß die Invarianz
von ds die Kovarianz des betreffenden Gleichungssystems zur Folge hat.

Bei der Verfolgung dieser allgemeinen Aufgaben stoßen wir aber
zunächst auf eine prinzipielle Schwierigkeit. Wir wissen nicht, bezüglich welcher Gruppe von Transformationen die gesuchten Gleichungen kovariant sein müssen. Am natürlichsten erscheint es zunächst, zu verlangen, daß
die Gleichungssysteme beliebigen Transformationen gegenüber kovariant sein sollen. Dem steht aber entgegen, daß die von uns aufgestellten
Gleichungen des Gravitationsfeldes diese Eigenschaft nicht besitzen. Wir
haben für die Gravitationsgleichungen nur beweisen können, daß sie
beliebigen linearen Transformationen gegenüber kovariant sind; wir
wissen aber nicht, ob es eine allgemeine Transformationsgruppe gibt,
der gegenüber die Gleichungen kovariant sind. Die Frage nach der
Existenz einer derartigen Gruppe für das Gleichungssystem (18) bzw. (21)
ist die wichtigste, welche sich an die hier gegebenen Ausführungen an-
knüpft. Jedenfalls sind wir bei dem gegenwärtigen Stande der Theorie
nicht berechtigt, die Kovarianz physikalischer Gleichungen beliebigen
Substitutionen gegenüber zu fordern.

Anderseits aber haben wir gesehen, daß sich eine Energie-Impuls-
Bilanzgleichung für materielle Vorgänge hat aufstellen lassen (§ 4, Gleichung 10), welche beliebige Transformationen gestattet. Es scheint deshalb doch natürlich, wenn wir voraussetzen, daß alle physikalischen
Gleichungssysteme mit Ausschluß der Gravitationsgleichungen so zu for-
mulieren sind, daß sie beliebigen Substitutionen gegenüber kovariant sind.
Die diesbezügliche Ausnahmestellung der Gravitationsgleichungen gegenüber allen anderen Systemen hängt nach meiner Meinung damit zusammen, daß nur erstere zweite Ableitungen der Komponenten des Fundamentaltensors enthalten dürften.

Die Aufstellung derartiger Gleichungssysteme erfordert die Hilfsmittel der verallgemeinerten Vektoranalysis, wie sie im II. Teil dargestellt ist.

Wir beschränksn uns hier darauf, anzugeben, wie man auf diesem Wege die elektromagnetischen Feldgleichungen für das Vakuum gewinnt.[1]) Wir gehen davon aus, daß die elektrische Ladung als etwas unveränderliches anzusehen ist. Ein unendlich kleiner, beliebig bewegter Körper habe die Ladung e und für einen mitbewegten Körper das Volumen dV_0 (Ruhvolumen). Wir definieren $\frac{e}{dV_0} = \varrho_0$ als die wahre Dichte der Elektrizität; diese ist ihrer Definition nach ein Skalar. Es ist daher

$$\varrho_0 \frac{dx_\nu}{ds} \qquad (\nu = 1, 2, 3, 4)$$

ein kontravarianter Vierervektor, den wir umformen, indem wir die Dichte ϱ der Elektrizität, aufs Koordinatensystem bezogen, durch die Gleichung

$$\varrho_0 dV_0 = \varrho dV$$

definieren. Unter Benutzung der Gleichung

$$dV_0 ds = \sqrt{-g} \cdot dV \cdot dt$$

des § 4 erhält man

$$\varrho_0 \frac{dx_\nu}{ds} = \frac{1}{\sqrt{-g}} \varrho \frac{dx_\nu}{dt},$$

d. h. den kontravarianten Vektor der elektrischen Strömung.

Das elektromagnetische Feld führen wir zurück auf einen speziellen, kontravarianten Tensor zweiten Ranges $\varphi_{\mu\nu}$ (einen Sechservektor) und bilden den „dualen" kontravarianten Tensor zweiten Ranges $\varphi_{\mu\nu}^*$ nach der Methode, die im II. Teil, § 3, auseinandergesetzt ist (Formel 42). Die Divergenz eines speziellen kontravarianten Tensors zweiten Ranges ist nach Formel 40 des II. Teiles, § 3

$$\frac{1}{\sqrt{-g}} \sum_\nu \frac{\partial}{\partial x_\nu} (\sqrt{-g} \cdot \varphi_{\mu\nu}).$$

Als Verallgemeinerung der Maxwell-Lorentzschen Feldgleichungen setzen wir die Gleichungen an

(23) $$\sum_\nu \frac{\partial}{\partial x_\nu} (\sqrt{-g} \cdot \varphi_{\mu\nu}) = \varrho \frac{dx_\mu}{dt}, \qquad (dt = dx_4)$$

(24) $$\sum_\nu \frac{\partial}{\partial x_\nu} (\sqrt{-g} \cdot \varphi_{\mu\nu}^*) = 0,$$

deren Kovarianz demnach evident ist. Setzen wir

$$\sqrt{-g} \cdot \varphi_{23} = \mathfrak{H}_x, \quad \sqrt{-g} \cdot \varphi_{31} = \mathfrak{H}_y, \quad \sqrt{-g} \cdot \varphi_{12} = \mathfrak{H}_z;$$
$$\sqrt{-g} \cdot \varphi_{14} = -\mathfrak{E}_x, \quad \sqrt{-g} \cdot \varphi_{24} = -\mathfrak{E}_y, \quad \sqrt{-g} \cdot \varphi_{34} = -\mathfrak{E}_z,$$

---

1) Vgl. hierzu auch die auf S. 23 zitierte Abhandlung von Kottler, § 3.

und
$$\varrho \frac{d x_\mu}{dt} = u_\mu,$$

so nimmt das Gleichungssystem (23) in ausführlicher Schreibweise die Form an

$$\frac{\partial \mathfrak{H}_s}{\partial y} - \frac{\partial \mathfrak{H}_y}{\partial z} - \frac{\partial \mathfrak{E}_x}{dt} = u_x$$

. . . .

$$\frac{\partial \mathfrak{E}_x}{\partial x} + \frac{\partial \mathfrak{E}_y}{\partial y} + \frac{\partial \mathfrak{E}_s}{\partial z} = \varrho,$$

welche Gleichungen bis auf die Wahl der Einheiten mit dem ersten Maxwellschen System übereinstimmen. Für die Bildung des zweiten Systems ist zunächst zu beachten, daß zu den Komponenten

$$\mathfrak{H}_x, \mathfrak{H}_y, \mathfrak{H}_z, -\mathfrak{E}_x, -\mathfrak{E}_y, -\mathfrak{E}_z$$

von
$$\sqrt{-g} \cdot \varphi_{\mu\nu}$$

die Komponenten
$$-\mathfrak{E}_x, -\mathfrak{E}_y, -\mathfrak{E}_z, \mathfrak{H}_x, \mathfrak{H}_y, \mathfrak{H}_z$$

der Ergänzung $f_{\mu\nu}$ gehören (II. Teil, § 3, Formeln 41a). Für den Fall des Fehlens des Gravitationsfeldes ergibt sich hieraus das zweite System, d. h. Gleichung (24) in der Form

$$-\frac{\partial \mathfrak{E}_x}{\partial x} + \frac{\partial \mathfrak{E}_y}{\partial z} - \frac{1}{c^2} \frac{\partial \mathfrak{H}_x}{\partial t} = 0$$

. . . .

. . . .

$$-\frac{1}{c^2} \frac{\partial \mathfrak{H}_x}{\partial x} - \frac{1}{c^2} \frac{\partial \mathfrak{H}_y}{\partial t} - \frac{1}{c^2} \frac{\partial \mathfrak{H}_z}{\partial z} = 0.$$

Damit ist erwiesen, daß die aufgestellten Gleichungen wirklich eine Verallgemeinerung derjenigen der gewöhnlichen Relativitätstheorie bilden.

§ 7. Kann das Gravitationsfeld auf einen Skalar zurückgeführt werden?

Bei der unleugbaren Kompliziertheit der hier vertretenen Theorie der Gravitation müssen wir uns ernstlich fragen, ob nicht die bisher ausschließlich vertretene Auffassung, nach welcher das Gravitationsfeld auf einen Skalar Φ zurückgeführt wird, die einzig naheliegende und berechtigte sei. Ich will kurz darlegen, warum wir diese Frage verneinen zu müssen glauben.

Es bietet sich bei Charakterisierung des Gravitationsfeldes durch einen Skalar ein Weg dar, welcher dem im Vorhergehenden eingeschlagenen ganz analog ist. Man setzt als Bewegungsgleichung des materiellen Punktes in Hamiltonscher Form an

$$\delta \left\{ \int \Phi ds \right\} = 0,$$

wobei ds das vierdimensionale Linienelement der gewöhnlichen Relativitätstheorie und Φ ein Skalar ist, und geht dann ganz analog vor wie im Vorhergehenden, ohne die gewöhnliche Relativitätstheorie verlassen zu müssen.

Auch hier ist der materielle Vorgang beliebiger Art durch einen Spannungs-Energie-Tensor $T_{\mu\nu}$ charakterisiert. Aber es ist bei dieser Auffassung ein Skalar maßgebend für die Wechselwirkung zwischen Gravitationsfeld und materiellem Vorgang. Dieser Skalar kann, worauf mich Herr Laue aufmerksam machte, nur

$$\sum_{\mu} T_{\mu\mu} = P$$

sein, den ich als den „Laueschen Skalar" bezeichnen will[1]). Dann kann man dem Satz von der Äquivalenz der trägen und der schweren Masse auch hier bis zu einem gewissen Grade gerecht werden. Herr Laue wies mich nämlich darauf hin, daß für ein abgeschlossenes System

$$\int P\,dV = \int T_{44}\,d\tau$$

ist. Hieraus ersieht man, das für die Schwere eines abgeschlossenen Systems auch nach dieser Auffassung seine Gesamtenergie maßgebend ist.

Die Schwere nicht abgeschlossener Systeme würde aber von den orthogonalen Spannungen T_{11} usw. abhängen, denen das System unterworfen ist. Daraus entstehen Konsequenzen, die mir unannehmbar erscheinen, wie an dem Beispiel der Hohlraumstrahlung gezeigt werden soll.

Für die Strahlung im Vakuum verschwindet bekanntlich der Skalar P. Ist die Strahlung in einem masselosen spiegelnden Kasten eingeschlossen, so erfahren deren Wände Zugspannungen, die bewirken, daß dem System, — als Ganzes genommen — eine schwere Masse $\int P\,d\tau$ zukommt, die der Energie E der Strahlung entspricht.

Statt nun aber die Strahlung in einen Hohlkasten einzuschließen, denke ich mir dieselbe begrenzt
S 1. durch die spiegelnden Wände eines festangeordneten Schachtes S,
2. durch zwei vertikal verschiebbare spiegelnde Wände W_1 und W_2, welche durch einen Stab fest miteinander verbunden sind.

In diesem Falle beträgt die schwere Masse $\int P\,d\tau$ des beweglichen Systems nur den dritten Teil des Wertes, der bei einem als Ganzes beweglichen Kasten auftritt. Man würde also zum Emporheben der Strah-

1) Vgl. II. Teil, § 1, letzte Formel.

lung entgegen einem Schwerefelde nur den dritten Teil der Arbeit aufwenden müssen als in dem vorhin betrachteten Falle, daß die Strahlung in einem Kasten eingeschlossen ist. Dies erscheint mir unannehmbar.

Ich muß freilich zugeben, daß für mich das wirksamste Argument dafür, daß eine derartige Theorie zu verwerfen sei, auf der Überzeugung beruht, daß die Relativität nicht nur orthogonalen linearen Substitutionen gegenüber besteht, sondern einer viel weiteren Substitutionsgruppe gegenüber. Aber wir sind schon deshalb nicht berechtigt, dieses Argument geltend zu machen, weil wir nicht imstande waren, die (allgemeinste) Substitutionsgruppe ausfindig zu machen, welche zu unseren Gravitationsgleichungen gehört.

II.

Mathematischer Teil.

Von Marcel Grossmann.

Die mathematischen Hilfsmittel für die Entwicklung der Vektoranalysis eines Gravitationsfeldes, das durch die Invarianz des Linienelementes

$$ds^2 = \sum_{\mu\nu} g_{\mu\nu}\, dx_\mu dx_\nu,$$

charakterisiert ist, gehen zurück auf die fundamentale Abhandlung von Christoffel[1]) über die Transformation der quadratischen Differentialformen. Ricci und Levi-Cività[2]) haben, ausgehend von den Christoffelschen Resultaten, ihre Methoden der absoluten, d. h. vom Koordinatensystem unabhängigen Differentialrechnung entwickelt, die gestatten, den Differentialgleichungen der mathematischen Physik eine invariante Form zu geben. Da aber die Vektoranalysis des auf beliebige krummlinige Koordinaten bezogenen euklidischen Raumes formal identisch ist mit der Vektoranalysis einer beliebigen, durch ihr Linienelement gegebenen Mannigfaltigkeit, so bietet es keine Schwierigkeiten, die vektoranalytischen Begriffsbildungen, wie sie in den letzten Jahren von Minkowski, Sommerfeld, Laue u. a. für die Relativitätstheorie entwickelt worden sind, auszudehnen auf die vorstehende allgemeine Theorie von Einstein.

Die allgemeine Vektoranalysis, die man so erhält, erweist sich bei einiger Übung als ebenso einfach zu handhaben, wie die spezielle des drei- oder vierdimensionalen euklidischen Raumes; ja die größere Allgemeinheit ihrer Begriffsbildungen verleiht ihr eine Übersichtlichkeit, die dem Spezialfall häufig genug abgeht.

Die Theorie der speziellen Tensoren (§ 3) ist in einer während des Entstehens dieser Arbeit erschienenen Abhandlung von Kottler[3])

---

1) Christoffel, Über die Transformation der homogenen Differentialausdrücke zweiten Grades, J. f. Math. 70 (1869), S. 46.

2) Ricci et Levi-Cività, Méthodes de calcul différentiel absolu et leurs applications, Math. Ann. 54 (1901), S. 125.

3) Kottler, Über die Raumzeitlinien der Minkowskischen Welt, Wien. Ber. 121 (1912).

vollständig behandelt worden und zwar, was im allgemeinen Falle nicht möglich ist, auf Grund der Theorie der Integralformen.

Da sich an die Gravitationstheorie von Einstein, insbesondere aber an das Problem der Differentialgleichungen des Gravitationsfeldes, eingehendere mathematische Untersuchungen werden knüpfen müssen, mag eine systematische Darstellung der allgemeinen Vektoranalysis am Platze sein. Dabei habe ich mit Absicht geometrische Hilfsmittel beiseite gelassen, da sie meines Erachtens wenig zur Veranschaulichung der Begriffsbildungen der Vektoranalysis beitragen.

§ 1. Allgemeine Tensoren.

Es sei

$$(1) \qquad ds^2 = \sum_{\mu\nu} g_{\mu\nu} dx_\mu dx_\nu$$

das Quadrat des Linienelementes, welches als invariantes Maß des Abstandes zweier unendlich-benachbarter Raum-Zeitpunkte betrachtet wird. Die folgenden Entwicklungen sind, so weit keine andere Bemerkung gemacht wird, von der Anzahl der Variabeln unabhängig; diese möge mit n bezeichnet sein.

Bei einer Transformation

$$(2) \qquad x_i = x_i(x_1', x_2', \ldots x_n') \qquad {\scriptstyle (i\,=\,1,\,2,\,\ldots\,n)}$$

der Variabeln, oder einer Transformation

$$(3) \qquad \begin{cases} dx_i = \sum_k \dfrac{\partial x_i}{\partial x_k'} dx_k' = \sum_k p_{ik} dx_k' \\[2mm] dx_i' = \sum_k \dfrac{\partial x_i'}{\partial x_k} dx_k = \sum_k \pi_{ki} dx_k \end{cases}$$

ihrer Differentiale, transformieren sich die Koeffizienten des Linienelementes gemäß der Formeln

$$(4) \qquad g_{rs}' = \sum_{\mu\nu} p_{\mu r} p_{\nu s} g_{\mu\nu}.$$

Es sei g die Diskriminante der Differentialform (1), d. h. die Determinante

$$g = |g_{\mu\nu}|.$$

Ist $\gamma_{\mu\nu}$ die durch die Diskriminante dividierte („normierte"), dem Element $g_{\mu\nu}$ adjungierte Unterdeterminante von g, so transformieren sich diese Größen $\gamma_{\mu\nu}$ nach den Formeln

$$(5) \qquad \gamma_{rs}' = \sum_{\mu\nu} \pi_{\mu r} \pi_{\nu s} \gamma_{\mu\nu}.$$

Wir definieren nun:

I. Der Inbegriff eines Systems von Funktionen $T_{i_1 i_2 \cdots i_\lambda}$ der Variabeln x heiße ein **kovarianter Tensor vom Range** λ, wenn diese Größen sich transformieren gemäß den Formeln

$$(6) \qquad T'_{r_1 r_2 \cdots r_\lambda} = \sum_{i_1 i_2 \cdots i_\lambda} p_{i_1 r_1} p_{i_2 r_2} \cdots p_{i_\lambda r_\lambda} \cdot T_{i_1 i_2 \cdots i_\lambda}.$$

II. Der Inbegriff eines Systems von Funktionen $\Theta_{i_1 i_2 \cdots i_\lambda}$ der Variabeln x heiße ein **kontravarianter Tensor vom Range** λ, wenn diese Größen sich transformieren gemäß den Formeln

$$(7) \qquad \Theta'_{r_1 r_2 \cdots r_\lambda} = \sum_{i_1 i_2 \cdots i_\lambda} \pi_{i_1 r_1} \pi_{i_2 r_2} \cdots \pi_{i_\lambda r_\lambda} \cdot \Theta_{i_1 i_2 \cdots i_\lambda}. \,^{[1]})$$

III. Der Inbegriff eines Systems von Funktionen $\mathfrak{T}_{i_1 i_2 \cdots i_\mu / k_1 k_2 \cdots k_\nu}$ der Variabeln x heiße ein **gemischter Tensor**, kovariant vom Range μ, kontravariant vom Range ν, wenn diese Größen sich transformieren nach den Formeln

$$(8)\ \mathfrak{T}'_{r_1 r_2 \cdots r_\mu / s_1 s_1 \cdots s_\nu} = \sum_{\substack{i_1 i_2 \cdots i_\mu \\ k_1 k_2 \cdots k_\nu}} p_{i_1 r_1} p_{i_2 r_2} \cdots p_{i_\mu r_\mu} \cdot \pi_{k_1 s_1} \pi_{k_1 s_2} \cdots \pi_{k_\nu s_\nu} \cdot \mathfrak{T}_{i_1 i_2 \cdots i_\mu / k_1 k_2 \cdots k_\nu}.$$

Aus diesen Definitionen und den Gleichungen (4) und (5) folgt:

Die Größen $g_{\mu\nu}$ bilden einen kovarianten, die Größen $\gamma_{\mu\nu}$ einen kontravarianten Tensor zweiten Ranges, die Fundamentaltensoren des Gravitationsfeldes im Falle $n = 4$.

Die Größen dx_i bilden nach Gleichung (3) einen kontravarianten Tensor ersten Ranges. Tensoren ersten Ranges nennt man auch **Vektoren erster Art** oder **Vierervektoren** bei $n = 4$.

Unmittelbar aus der Definition der Tensoren ergeben sich die folgenden algebraischen Tensoroperationen:

1. **Die Summe zweier gleichartiger Tensoren vom Range** λ **ist wieder ein gleichartiger Tensor vom Range** λ, dessen Komponenten durch Addition der entsprechenden Komponenten beider Tensoren entstehen.

---

1) Unsere kovarianten (kontravarianten) Tensoren vom Range λ sind also identisch mit den „kovarianten (kontravarianten) Systemen λ^{ter} Ordnung" von Ricci und Levi-Cività und werden von diesen Autoren bezeichnet mit $X_{r_1 r_2 \cdots r_\lambda}$ bzw. $X^{r_1 r_2 \cdots r_\lambda}$. So viele Vorteile diese letztere Bezeichnung auch bietet, so haben uns doch Komplikationen in zusammengesetzteren Gleichungen gezwungen, die obigen Bezeichnungen zu wählen, also kovariante Tensoren mit lateinischen, kontravariante mit griechischen, gemischte mit deutschen Buchstaben zu bezeichnen. Kovariante und kontravariante Tensoren sind besondere Fälle der gemischten Tensoren.

2. Das äußere Produkt zweier kovarianter (kontravarianter) Tensoren vom Range λ bzw. μ ist ein kovarianter (kontravarianter) Tensor vom Range $\lambda + \mu$ mit den Komponenten

$$(9) \qquad T_{i_1 i_2 \cdots i_\lambda k_1 k_2 \cdots k_\mu} = A_{i_1 i_2 \cdots i_\lambda} \cdot B_{k_1 k_2 \cdots k_\mu},$$

bzw.

$$(9') \qquad \Theta_{i_1 i_2 \cdots i_\lambda k_1 k_2 \cdots k_\mu} = \Phi_{i_1 i_2 \cdots i_\lambda} \cdot \Psi_{k_1 k_2 \cdots k_\mu}.$$

3. Als inneres Produkt zweier Tensoren bezeichnen wir

a) den kovarianten Tensor

$$(10) \qquad T_{i_1 i_2 \cdots i_\lambda} = \sum_{k_1 k_2 \cdots k_\mu} \Phi_{k_1 k_2 \cdots k_\mu} \cdot A_{i_1 i_2 \cdots i_\lambda k_1 k_2 \cdots k_\mu},$$

b) den kontravarianten Tensor

$$(11) \qquad \Theta_{i_1 i_2 \cdots i_\lambda} = \sum_{k_1 k_2 \cdots k_\mu} A_{k_1 k_2 \cdots k_\mu} \cdot \Phi_{i_1 i_2 \cdots i_\lambda k_1 k_2 \cdots k_\mu},$$

c) den gemischten Tensor

$$(12) \qquad \mathfrak{T}_{r_1 r_2 \cdots r_\mu s_1 s_2 \cdots s_\nu} = \sum_{k_1 k_2 \cdots k_\lambda} A_{k_1 k_2 \cdots k_\lambda r_1 r_2 \cdots r_\mu} \cdot \Phi_{k_1 k_2 \cdots k_\lambda s_1 s_2 \cdots s_\nu},$$

oder ganz allgemein, die drei Fälle a) bis c) mit enthaltend

$$d) \quad \mathfrak{T}_{r_1 r_2 \cdots r_\mu u_1 u_2 \cdots u_\alpha s_1 s_2 \cdots s_\nu t_1 t_2 \cdots t_\beta} = \sum_{k_1 k_2 \cdots k_\lambda} \mathfrak{A}_{r_1 r_2 \cdots r_\mu | k_1 k_2 \cdots k_\lambda s_1 s_2 \cdots s_\nu} \cdot \mathfrak{B}_{k_1 k_2 \cdots k_\lambda u_1 u_2 \cdots u_\alpha t_1 t_2 \cdots t_\beta}.$$

Die der gewöhnlichen Vektoranalysis entnommenen Bezeichnungen „äußeres und inneres Produkt" rechtfertigen sich, weil jene Operationen sich letzten Endes als besondere Fälle der hier betrachteten ergeben.

Ist in den Fällen a) oder b) der Rang λ gleich Null, so ist das innere Produkt ein Skalar.

4. Reziprozität eines kovarianten und eines kontravarianten Tensors. Aus einem kovarianten Tensor vom Range λ bildet man den reziproken kontravarianten Tensor vom Range λ durch λ-fache innere Multiplikation mit dem kontravarianten Fundamentaltensor:

$$(13) \qquad \Theta_{i_1 i_2 \cdots i_\lambda} = \sum_{k_1 k_2 \cdots k_\lambda} \gamma_{i_1 k_1} \gamma_{i_2 k_2} \cdots \gamma_{i_\lambda k_\lambda} \cdot T_{k_1 k_2 \cdots k_\lambda},$$

woraus durch Auflösung

$$(14) \qquad T_{i_1 i_2 \cdots i_\lambda} = \sum_{k_1 k_2 \cdots k_\lambda} g_{i_1 k_1} g_{i_2 k_2} \cdots g_{i_\lambda k_\lambda} \cdot \Theta_{k_1 k_2 \cdots k_\lambda}.$$

Man findet daher aus einem Tensor einen Skalar, in dem man ihn mit seinem reziproken Tensor multipliziert nach der Formel

$$(15) \qquad \sum_{i_1 i_2 \cdots i_\lambda} T_{i_1 i_2 \cdots i_\lambda} \cdot \Theta_{i_1 i_2 \cdots i_\lambda}$$

Ein kovarianter (kontravarianter) Tensor ersten Ranges (Vierervektor bei $n = 4$) hat die Invariante

$$\sum_{ik} \gamma_{ik} T_i T_k$$

beziehungsweise

$$\sum_{ik} g_{ik} \Theta_i \Theta_k.$$

In der gewöhnlichen Relativitätstheorie ist die Kontravarianz identisch der Kovarianz und obige Invariante wird zum Quadrat des Betrages des Vierervektors

$$T_x^2 + T_y^2 + T_z^2 + T_l^2.$$

Ein kovarianter (kontravarianter) Tensor zweiten Ranges hat die Invariante

$$\sum_{ik} \gamma_{ik} T_{ik}$$

beziehungsweise

$$\sum_{ik} g_{ik} \Theta_{ik},$$

die im Falle der bisherigen Relativitätstheorie zu

$$T_{xx} + T_{yy} + T_{zz} + T_{ll}$$

wird.[1])

§ 2. Differentialoperationen an Tensoren.

Wir führen folgende allgemeine Definitionen ein:

I. Als **Erweiterung** eines kovarianten (kontravarianten) Tensors vom Range λ bezeichnen wir den kovarianten (kontravarianten) Tensor vom Range $\lambda + 1$, der durch „kovariante (kontravariante) Differentiation" aus jenem hervorgeht.

Nach Christoffel (l. c.) ist

$$(16) \qquad T_{r_1 r_2 \cdots r_\lambda s} = \frac{\partial T_{r_1 r_2 \cdots r_\lambda}}{\partial x_s} -$$
$$- \sum_k \left(\begin{Bmatrix} r_1 s \\ k \end{Bmatrix} T_{k r_2 \cdots r_\lambda} + \begin{Bmatrix} r_2 s \\ k \end{Bmatrix} T_{r_1 k \cdots r_\lambda} + \cdots + \begin{Bmatrix} r_\lambda s \\ k \end{Bmatrix} T_{r_1 r_2 \cdots k} \right)$$

---

1) Wir verzichten im folgenden darauf, jeweilen die besondere Form anzugeben, welche unsere Formeln im Falle der gewöhnlichen Relativitätstheorie annehmen, begnügen uns vielmehr damit, hinzuweisen auf die nachstehenden Darstellungen:

1. Minkowski, Die Grundgleichungen für die elektromagnetischen Vorgänge in bewegten Körpern, Göttinger Nachrichten 1908.

2. Sommerfeld, Zur Relativitätstheorie I und II, Ann. d. Physik, vierte Folge, 32 (1910) und 33 (1910).

3. Laue, Das Relativitätsprinzip. Die Wissenschaft, Heft 38, 2. A. (1913).

ein kovarianter Tensor vom Range $\lambda + 1$, der aus dem kovarianten Tensor vom Range λ hervorgeht. Ricci und Levi-Cività nennen die Differentialoperation der rechten Seite dieser Gleichung die „kovariante Differentiation" des Tensors $T_{r_1 r_2 \cdots r_\lambda}$. Hierbei bedeutet

$$(17) \qquad \left\{ \begin{matrix} r\,s \\ u \end{matrix} \right\} = \sum_t \gamma_{ut} \left[\begin{matrix} r\,s \\ t \end{matrix} \right],$$

$$(18) \qquad \left[\begin{matrix} r\,s \\ t \end{matrix} \right] = \tfrac{1}{2} \left(\frac{\partial g_{rt}}{\partial x_s} + \frac{\partial g_{st}}{\partial x_r} - \frac{\partial g_{rs}}{\partial x_t} \right).$$

$\left[\begin{matrix} r\,s \\ t \end{matrix} \right]$ und $\left\{ \begin{matrix} r\,s \\ u \end{matrix} \right\}$ sind die Christoffelschen Drei-Indizes-Symbole erster bzw. zweiter Art; durch Auflösung der Gleichungen (17) findet man

$$(19) \qquad \left[\begin{matrix} r\,s \\ u \end{matrix} \right] = \sum_t g_{ut} \left\{ \begin{matrix} r\,s \\ t \end{matrix} \right\}. \text{[1]}$$

Führt man in die Gleichung (16) an Stelle der kovarianten Tensoren die zu ihnen reziproken kontravarianten Tensoren ein, so erhält man als „kontravariante Erweiterung"

$$(20) \quad \Theta_{r_1 r_2 \cdots r_\lambda s} = \sum_{ik} \gamma_{si} \left(\frac{\partial \Theta_{r_1 r_2 \cdots r_\lambda}}{\partial x_i} + \left\{ \begin{matrix} i\,k \\ r_1 \end{matrix} \right\} \Theta_{k r_2 \cdots r_\lambda} + \left\{ \begin{matrix} i\,k \\ r_2 \end{matrix} \right\} \Theta_{r_1 k \cdots r_\lambda} + \cdots + \left\{ \begin{matrix} i\,k \\ r_\lambda \end{matrix} \right\} \Theta_{r_1 r_2 \cdots k} \right).$$

II. Als **Divergenz** eines kovarianten (kontravarianten) Tensors vom Range λ bezeichnen wir den kovarianten (kontravarianten) Tensor vom Range $\lambda - 1$, der durch innere Multiplikation der Erweiterung mit dem kontravarianten (kovarianten) Fundamentaltensor entsteht.

Somit ist die Divergenz des kovarianten Tensors $T_{r_1 r_2 \cdots r_\lambda}$ der Tensor

$$(21) \qquad T_{r_2 r_3 \cdots r_\lambda} = \sum_{s r_1} \gamma_{s r_1} T_{r_1 \cdots r_\lambda s},$$

und die Divergenz des kontravarianten Tensors $\Theta_{r_1 r_2 \cdots r_\lambda}$ ist der Tensor

$$(22) \qquad \Theta_{r_2 r_3 \cdots r_\lambda} = \sum_{s r_1} g_{s r_1} \Theta_{r_1 \cdots r_\lambda s}.$$

Die Divergenz eines Tensors geht nicht eindeutig aus diesem hervor; das Resultat ändert sich im allgemeinen, wenn man in den Gleichungen (21) und (22) r_1 durch einen der Indizes $r_2, r_3 \ldots r_\lambda$ ersetzt.

III. Als **verallgemeinerte Laplacesche Operation** an einem Tensor bezeichnen wir die Aufeinanderfolge der Erweiterung und der Divergenz. Die verallgemeinerte Laplacesche Operation läßt daher aus einem Tensor einen gleichartigen gleichen Ranges hervorgehen.

Von besonderem Interesse sind die Fälle $\lambda = 0, 1, 2$.

---

[1] Auf Grund dieser Formeln beweist man leicht, daß die Erweiterung des Fundamentaltensors identisch verschwindet.

a) $\lambda = 0$.

Der Ausgangstensor ist ein Skalar T, den wir als ko- oder kontravarianten Tensor vom Range 0 betrachten können.

$$(23) \qquad T_r = \frac{\partial T}{\partial x_r}$$

ist die kovariante Erweiterung des Skalars T, d. i. ein kovarianter Tensor ersten Ranges (kovarianter Vierervektor für $n = 4$), den man den Gradienten des Skalars nennt. Die Invariante

$$(24) \qquad \sum_{rs} \gamma_{rs} \frac{\partial T}{\partial x_r} \frac{\partial T}{\partial x_s}$$

ist der erste Beltramische Differentialparameter des Skalars T.

Um die Divergenz des Gradienten zu bilden, hat man aus seiner Erweiterung

$$T_{rs} = \frac{\partial^2 T}{\partial x_r \partial x_s} - \sum_k \begin{Bmatrix} rs \\ k \end{Bmatrix} \frac{\partial T}{\partial x_k}$$

den Skalar

$$\sum_{rs} \gamma_{rs} T_{rs}$$

zu bilden, dem man die Form

$$(25) \qquad \frac{1}{\sqrt{g}} \sum_{rs} \frac{\partial}{\partial x_s} \left(\sqrt{g}\, \gamma_{rs} \frac{\partial T}{\partial x_r} \right)$$

geben kann.[1]) Die Divergenz des Gradienten ist das Resultat der verallgemeinerten Laplaceschen Operation ausgeführt am Skalar T und ist identisch mit dem zweiten Beltramischen Differentialparameter des Skalars T.

b) $\lambda = 1$.

Der Ausgangstensor sei ein kovarianter Vierervektor, könnte aber ebensogut ein kontravarianter Vierervektor sein.

Die kovariante Erweiterung ist nach (16)

$$(26) \qquad T_{rs} = \frac{\partial T_r}{\partial x_s} - \sum_k \begin{Bmatrix} rs \\ k \end{Bmatrix} T_k .$$

Die Divergenz ist

$$(27) \qquad \sum_{rs} \gamma_{rs} T_{rs} = \sum_{rsk} \gamma_{rs} \left(\frac{\partial T_r}{\partial x_s} - \begin{Bmatrix} rs \\ k \end{Bmatrix} T_k \right),$$

der wir nach (17) die Form geben:

$$(28) \sum_{rs} \gamma_{rs} T_{rs} = \sum_{rskl} \left(\frac{\partial}{\partial x_s} (\gamma_{rs} T_r) - \frac{\partial \gamma_{rs}}{\partial x_s} \cdot T_r - \frac{1}{2} \gamma_{rs} \gamma_{kl} \left(\frac{\partial g_{rl}}{\partial x_s} + \frac{\partial g_{sl}}{\partial x_r} - \frac{\partial g_{rs}}{\partial x_l} \right) T_k \right)$$

---

1) Siehe z. B. Bianchi-Lukat, Vorlesungen über Differentialgeometrie, erste Auflage, S. 47; oder auch die Umrechnung der Divergenz eines Vierervektors im nachstehenden Falle b).

Eliminiert man $\dfrac{\partial \gamma_{rs}}{\partial x_s}$ vermöge der Formel[1])

(29)
$$\frac{\partial \gamma_{rs}}{\partial x_t} = -\sum_{\varrho\sigma} \gamma_{r\varrho}\gamma_{s\sigma}\frac{\partial g_{\varrho\sigma}}{\partial x_t},$$

so heben sich in Gleichung (28) die drei mittleren Glieder unter dem Summenzeichen auf und es bleibt neben dem ersten Gliede

$$\sum_{rskl} \tfrac{1}{2}\gamma_{rs}\frac{\partial g_{rs}}{\partial x_t}\cdot\gamma_{kl}\,T_k = \sum_{kl}\gamma_{kl}\,T_k\frac{\partial \log\sqrt{g}}{\partial x_t},$$

so daß man für die Divergenz des kovarianten Vierervektors[2]) findet

(30)
$$\sum_{rs}\gamma_{rs}\,T_{rs} = \frac{1}{\sqrt{g}}\sum_{rs}\frac{\partial}{\partial x_s}\left(\sqrt{g}\,\gamma_{rs}\,T_r\right).$$

c) $\lambda = 2$.

Der Ausgangstensor sei ein kontravarianter Tensor zweiten Ranges Θ_{rs}, dessen Erweiterung nach Formel (20) lautet

(31)
$$\Theta_{rst} = \sum_{ik}\gamma_{ti}\left(\frac{\partial \Theta_{rs}}{\partial x_i} + \begin{Bmatrix} ik \\ r \end{Bmatrix}\Theta_{ks} + \begin{Bmatrix} ik \\ s \end{Bmatrix}\Theta_{rk}\right).$$

Hieraus ergibt sich als Divergenz des kontravarianten Tensors Θ_{rs} entweder die Zeilendivergenz

(32)
$$\Theta_r = \sum_{st} g_{st}\Theta_{rst} = \sum_{sk}\left(\frac{\partial \Theta_{rs}}{\partial x_s} + \begin{Bmatrix} sk \\ r \end{Bmatrix}\Theta_{ks} + \begin{Bmatrix} sk \\ s \end{Bmatrix}\Theta_{rk}\right),$$

oder die Kolonnendivergenz

(33)
$$\Theta_s = \sum_{rt} g_{rt}\Theta_{rst} = \sum_{rk}\left(\frac{\partial \Theta_{rs}}{\partial x_r} + \begin{Bmatrix} rk \\ r \end{Bmatrix}\Theta_{ks} + \begin{Bmatrix} rk \\ s \end{Bmatrix}\Theta_{rk}\right),$$

---

1) Diese Formel, die wir auch in § 4 bei der Aufstellung der Differential-gleichungen des Gravitationsfeldes verwenden, beweisen wir folgendermaßen:

Es ist
$$\sum_l g_{il}\gamma_{kl} = \delta_{ik}\ (0\ \text{oder}\ 1),$$

also
$$\sum_l g_{il}\frac{\partial \gamma_{kl}}{\partial x_t} = -\sum_l \gamma_{kl}\frac{\partial g_{il}}{\partial x_t},$$

wo t irgend eine der Zahlen $1, 2, \ldots n$ ist.

Für ein bestimmtes k erhält man so n Gleichungen $(i = 1, 2, \ldots n)$ mit den n Unbekannten $\dfrac{\partial \gamma_{kl}}{\partial x_t}$, $(l = 1, 2, \ldots n)$, deren Auflösung die Formel des Textes liefert.

2) Zu dem nämlichen Ergebnis gelangt Kottler (l. c. pag. 21) ausgehend von einem speziellen Tensor dritten Ranges (vgl. § 3 dieser Abhandlung) mit Hilfe der Theorie der Integralformen.

zwei Differentialoperationen, die für symmetrische Tensoren zusammen-
fallen. Weil

$$
(34) \qquad \sum_r \left\{ {r\,k \atop r} \right\} = \sum_{rs} \gamma_{rs} \left[{r\,k \atop s} \right] = \sum_{rs} \tfrac{1}{2} \gamma_{rs} \frac{\partial g_{rs}}{\partial x_k} = \frac{\partial \log \sqrt{g}}{\partial x_k}
$$

ist, so läßt sich die Formel (33) auch zusammenfassen in

$$
(35) \qquad \Theta_s = \frac{1}{\sqrt{g}} \sum_r \frac{\partial}{\partial x_r} \left(\sqrt{g} \cdot \Theta_{rs} \right) + \sum_{rk} \left\{ {r\,k \atop s} \right\} \Theta_{rk}.
$$

§ 3. Spezielle Tensoren (Vektoren).

Ein kovarianter (kontravarianter) Tensor heiße **speziell**, wenn
seine Komponenten ein System von **alternierenden Funktionen** der
Grundvariabeln bilden.

Die Komponenten eines speziellen Tensors sind demnach den folgenden Bedingungen unterworfen:

1. Es ist $T_{r_1 r_2 \cdots r_\lambda} = 0$, wenn zwei der Indizes r_1, r_2, ... r_λ einander gleich sind.

2. Unterscheiden sich r_1, r_2, ... r_λ und s_1, s_2, ... s_λ nur durch die
Reihenfolge der Indizes, so ist $T_{r_1 r_2 \cdots r_\lambda} = \pm\, T_{s_1 s_2 \cdots s_\lambda}$, je nachdem r_1, r_2,
... r_λ und s_1, s_2, ... s_λ Permutationen derselben Klasse sind oder nicht.
Zwei Permutationen gehören bekanntlich zu der gleichen Klasse, wenn
beide durch eine gerade bezw. ungerade Anzahl von bloßen Vertau-
schungen zweier Indizes aus der Grundpermutation 1, 2, ... n hervorgehen.

Die Anzahl der linear unabhängigen Komponenten eines speziellen
Tensors vom Range λ ist demnach $\binom{n}{\lambda}$.

Die Theorie der speziellen Tensoren gestaltet sich vermöge dieser
Eigenschaften einfacher, aber auch reichhaltiger als die der allgemeinen
Tensoren; sie ist von besonderer Bedeutung für die mathematische
Physik, weil die Theorie der **Vektoren** λ^{ter} Art (Vierer-, Sechservektoren bei $n = 4$) sich zurückführen läßt auf die **speziellen Tensoren
vom Range λ**. Vom Standpunkte der allgemeinen Theorie aus ist es
zweckmäßiger von den Tensoren auszugehen und die Vektoren lediglich
als spezielle Tensoren zu behandeln.

Wichtig für die Vektoranalysis der n-dimensionalen Mannigfaltigkeit

$$
ds^2 = \sum_{\mu\nu} g_{\mu\nu}\, dx_\mu\, dx_\nu
$$

ist ein spezieller Tensor n^{ten} Ranges, der mit der Diskriminante g des

Linienelementes verknüpft ist.[1]) Diese Diskriminante transformiert sich
ja gemäß der Gleichung

$$(36) \qquad\qquad g' = p^2 \cdot g,$$

wo

$$p = |\,p_{ik}\,| = \left|\frac{\partial x_i}{\partial x_k'}\right|$$

die Funktionaldeterminante der Substitution ist. Gibt man $\sqrt{g}$ für das
ursprüngliche Bezugssystem ein bestimmtes Vorzeichen, und setzt man
fest, daß sich dieses Vorzeichen bei einer Transformation ändern soll
oder nicht, je nachdem die Substitutionsdeterminante p negativ oder
positiv ist, so hat die Gleichung

$$(37) \qquad\qquad \sqrt{g'} = p \cdot \sqrt{g}$$

exakte Bedeutung mit Einschluß der Vorzeichen.

Es sei nun $\delta_{r_1 r_2 \cdots r_n}$ gleich Null, wenn zwei der Indizes einander
gleich sind, dagegen ± 1, wenn dies nicht der Fall ist und die Permu-
tation $r_1, r_2, \ldots r_n$ durch eine gerade bezw. ungerade Anzahl von Ver-
tauschungen zweier Indizes aus der Grundpermutation $1, 2, \ldots n$ hervorgeht.

Dann sind

$$(38) \qquad\qquad e_{r_1 r_2 \cdots r_n} = \delta_{r_1 r_2 \cdots r_n} \cdot \sqrt{g}$$

die Komponenten eines speziellen kovarianten Tensors n ten Ranges, den
wir den kovarianten Diskriminantentensor nennen wollen. Denn
eine Transformation liefert zunächst

$$e'_{r_1 r_2 \cdots r_n} = \delta_{r_1 r_2 \cdots r_n} \cdot \sqrt{g'} = \delta_{r_1 r_2 \cdots r_n} \cdot p \sqrt{g};$$

da aber

$$p = \sum_{i_1 i_2 \cdots i_n} \delta_{i_1 i_2 \cdots i_n} \, p_{i_1 1} p_{i_2 2} \cdots p_{i_n n} = \delta_{r_1 r_2 \cdots r_n} \cdot \sum_{i_1 i_2 \cdots i_n} \delta_{i_1 i_2 \cdots i_n} \cdot p_{i_1 r_1} p_{i_2 r_2} \cdots p_{i_n r_n}$$

ist, so folgt

$$e'_{r_1 r_2 \cdots r_n} = \sqrt{g} \cdot \sum_{i_1 i_2 \cdots i_n} \delta_{i_1 i_2 \cdots i_n} \cdot p_{i_1 r_1} p_{i_2 r_2} \cdots p_{i_n r_n},$$

also wegen der Definition (38)

$$e'_{r_1 r_2 \cdots r_n} = \sum_{i_1 i_2 \cdots i_n} e_{i_1 i_2 \cdots i_n} \cdot p_{i_1 r_1} p_{i_2 r_2} \cdots p_{i_n r_n}.$$

Für den reziproken kontravarianten Tensor findet man nach (13)

$$\varepsilon_{i_1 i_2 \cdots i_n} = \sum_{r_1 r_2 \cdots r_n} \gamma_{i_1 r_1} \gamma_{i_2 r_2} \cdots \gamma_{i_n r_n} \cdot e_{r_1 r_2 \cdots r_n},$$

$$\varepsilon_{i_1 i_2 \cdots i_n} = \sqrt{g} \cdot \sum_{r_1 r_2 \cdots r_n} \delta_{r_1 r_2 \cdots r_n} \cdot \gamma_{i_1 r_1} \gamma_{i_2 r_2} \cdots \gamma_{i_n r_n},$$

$$\varepsilon_{i_1 i_2 \cdots i_n} = \delta_{i_1 i_2 \cdots i_n} \cdot \sqrt{g} \cdot \sum_{r_1 r_2 \cdots r_n} \delta_{r_1 r_2 \cdots r_n} \cdot \gamma_{1 r_1} \gamma_{2 r_2} \cdots \gamma_{n r_n}.$$

---

1) Das „System ε“ von Ricci und Levi-Cività, l. c., pag. 135.

Da aber die Determinante der normierten Unterdeterminanten γ_{ik}

$$| \gamma_{ik} | = \frac{1}{g}$$

ist, so folgt

(39)

$$\varepsilon_{i_1 i_2 \cdots i_n} = \frac{\delta_{i_1 i_2 \cdots i_n}}{\sqrt{g}}.$$

Die Bedeutung des kovarianten (kontravarianten) Diskriminanten-tensors liegt darin, daß seine innere Multiplikation mit einem kontra-varianten (kovarianten) Tensor vom Range λ einen gleichartigen Tensor vom Range $\lambda - n$ liefert, wobei der Tensor von entgegengesetzter Art wird, wenn $\lambda - n$ negativ ist. (Ergänzung des Tensors.)

Wenn

$$n = 4$$

ist, so gibt es spezielle Tensoren bis zum vierten Rang, da alle spe-ziellen Tensoren höheren Ranges identisch verschwinden.

Die nichtverschwindenden Komponenten eines speziellen kovarian-ten Tensors vierten Ranges sind alle einander gleich oder entgegenge-setzt gleich. Die Ergänzung (innere Multiplikation mit dem kontra-varianten Diskriminantentensor) ergibt einen Skalar, so daß die Diffe-rentialoperationen, die an einem speziellen Tensor vierten Ranges aus-geführt werden können, damit zurückgeführt sind auf die Differential-operationen an einen Skalar.

Die Ergänzung eines speziellen kovarianten Tensors dritten Ranges ist ein kontravarianter Vektor erster Art.

Die Ergänzung eines speziellen kovarianten Tensors zweiten Ranges ist ein kontravarianter, spezieller Tensor zweiten Ranges.

Endlich führt die Ergänzung eines speziellen kovarianten Vektors erster Art auf einen kontravarianten Tensor dritten Ranges.

Die Untersuchung des Einflusses des Gravitationsfeldes auf die physikalischen Vorgänge (I. Teil, § 6) erfordert die eingehendere Be-handlung der speziellen Tensoren zweiten Ranges (Sechservektoren).

Ist $\Theta_{\mu\nu}$ ein spezieller Tensor zweiten Ranges, so reduziert sich seine Divergenz (Formel 35)

$$\Theta_\mu = \sum_\nu \frac{1}{\sqrt{g}} \frac{\partial}{\partial x_\nu} (\sqrt{g} \cdot \Theta_{\mu\nu}) + \sum_{\nu\varkappa} \left\{ \begin{matrix} \nu\varkappa \\ \mu \end{matrix} \right\} \Theta_{\nu\varkappa}$$

wegen

$$\Theta_{\nu\varkappa} = - \Theta_{\varkappa\nu}, \quad \Theta_{\nu\nu} = 0$$

auf

(40)

$$\Theta_\mu = \sum_\nu \frac{1}{\sqrt{g}} \frac{\partial}{\partial x_\nu} (\sqrt{g} \cdot \Theta_{\mu\nu}).$$

Wir leiten ferner aus einem kontravarianten Tensor zweiten Ranges $\Theta_{\mu\nu}$ folgendermaßen den **dualen** kontravarianten Tensor zweiten Ranges Θ_{rs}^* ab.

Wir bilden zuerst die Ergänzung[1])

$$(41) \qquad T_{ik} = \tfrac{1}{2} \sum_{\mu\nu} e_{ik\mu\nu} \cdot \Theta_{\mu\nu},$$

oder also

$$(41\,\mathrm{a}) \qquad \begin{cases} T_{12} = \sqrt{g} \cdot \Theta_{34}, & T_{13} = \sqrt{g} \cdot \Theta_{42}, & T_{14} = \sqrt{g} \cdot \Theta_{23}; \\ T_{23} = \sqrt{g} \cdot \Theta_{14}, & T_{24} = \sqrt{g} \cdot \Theta_{31}, & T_{34} = \sqrt{g} \cdot \Theta_{12}. \end{cases}$$

Der gesuchte duale Tensor ist nun reziprok zu dieser Ergänzung, lautet daher

$$(42) \qquad \Theta_{rs}^* = \sum_{ik} \gamma_{ir}\gamma_{ks} \cdot T_{ik} = \tfrac{1}{2} \sum_{ik\mu\nu} \gamma_{ir}\gamma_{ks} e_{ik\mu\nu} \cdot \Theta_{\mu\nu}.$$

Die Reihenfolge der beiden Operationen — Ergänzung und Bildung des reziproken Tensors — ist wegen der Reziprozität der beiden Diskriminantentensoren vertauschbar. —

§ 4. Mathematische Ergänzungen zum physikalischen Teil.

1. **Beweis der Kovarianz der Impuls-Energiegleichungen.**

Es ist zu beweisen, daß sich die Gleichungen (10) des I. Teiles, S. 10, die vom Faktor $\sqrt{-1}$ abgesehen lauten

$$\sum_{\mu\nu} \frac{\partial}{\partial x_\nu} \left(\sqrt{g} \cdot g_{\sigma\mu} \cdot \Theta_{\mu\nu} \right) - \tfrac{1}{2} \sqrt{g} \sum_{\mu\nu} \frac{\partial g_{\mu\nu}}{\partial x_\sigma} \cdot \Theta_{\mu\nu} = 0, \qquad (\sigma = 1,2,3,4)$$

beliebigen Transformationen gegenüber kovariant verhalten.

Nach Formel (35) ist die Divergenz des kontravarianten Tensors $\Theta_{\mu\nu}$

$$\Theta_\mu = \sum_\nu \frac{1}{\sqrt{g}} \frac{\partial}{\partial x_\nu} \left(\sqrt{g} \cdot \Theta_{\mu\nu} \right) + \sum_{\nu k} \left\{ \begin{matrix} \nu k \\ \mu \end{matrix} \right\} \Theta_{\nu k}.$$

Der zu diesem kontravarianten Vektor Θ_μ reziproke kovariante Vektor T_σ ist also

$$T_\sigma = \sum_\mu g_{\sigma\mu} \Theta_\mu = \sum_{\mu\nu k} \left(\frac{1}{\sqrt{g}} \frac{\partial}{\partial x_\nu} \left(\sqrt{g} \cdot g_{\sigma\mu} \cdot \Theta_{\mu\nu} \right) - \frac{\partial g_{\sigma\mu}}{\partial x_\nu} \cdot \Theta_{\mu\nu} + g_{\sigma\mu} \left\{ \begin{matrix} \nu k \\ \mu \end{matrix} \right\} \cdot \Theta_{\nu k} \right).$$

Das letzte Glied dieser Summe ist aber gleich

$$\sum_{\nu k} \left[\begin{matrix} \nu k \\ \sigma \end{matrix} \right] \Theta_{\nu k} = \sum_{\mu\nu} \tfrac{1}{2} \left(\frac{\partial g_{\mu\sigma}}{\partial x_\nu} + \frac{\partial g_{\nu\sigma}}{\partial x_\mu} - \frac{\partial g_{\mu\nu}}{\partial x_\sigma} \right) \cdot \Theta_{\mu\nu}.$$

---

[1]) Der Faktor $\tfrac{1}{2}$ dient zur Vereinfachung des Resultates, ohne invariantentheoretisch von Belang zu sein.

Also bleibt

$$T_\sigma = \sum_{\mu\nu} \frac{1}{\sqrt{g}} \frac{\partial}{\partial x_\nu} \left(\sqrt{g} \cdot g_{\sigma\mu} \Theta_{\mu\nu} \right) - \frac{1}{2} \sum_{\mu\nu} \frac{\partial g_{\mu\nu}}{\partial x_\sigma} \cdot \Theta_{\mu\nu},$$

d. h. bis auf den Faktor $\dfrac{1}{\sqrt{g}}$ die linke Seite der untersuchten Gleichung. Dividiert man also jene Gleichung durch $\sqrt{g}$, so stellt ihre linke Seite die σ-Komponente eines kovarianten Vektors dar, ist also in der Tat kovariant. Man kann daher den Inhalt jener vier Gleichungen auch so aussprechen:

Die Divergenz des (kontravarianten) Spannungs-Energie-tensors der materiellen Strömung bzw. des physikalischen Vorganges verschwindet.

2. Differentialtensoren einer durch ihr Linienelement gegebenen Mannigfaltigkeit.

Das Problem der Aufstellung der Differentialgleichungen eines Gravitationsfeldes (I. Teil, § 5) lenkt die Aufmerksamkeit auf die Differentialinvarianten und Differentialkovarianten der quadratischen Differentialform

$$ds^2 = \sum_{\mu\nu} g_{\mu\nu} dx_\mu dx_\nu.$$

Die Theorie dieser Differentialkovarianten führt im Sinne unserer allgemeinen Vektoranalysis auf die Differentialtensoren, die mit einem Gravitationsfeld gegeben sind. Das vollständige System dieser Differentialtensoren (beliebigen Transformationen gegenüber) geht zurück auf eine von Riemann[1]) und unabhängig von diesem von Christoffel[2]) gefundenen kovarianten Differentialtensor vierten Ranges, den wir den Riemannschen Differentialtensor nennen wollen und der folgendermaßen lautet:

$$(43) \quad \begin{aligned} R_{iklm} = (ik, lm) = \frac{1}{2} \left(\frac{\partial^2 g_{im}}{\partial x_k \partial x_l} + \frac{\partial^2 g_{kl}}{\partial x_i \partial x_m} - \frac{\partial^2 g_{il}}{\partial x_k \partial x_m} - \frac{\partial^2 g_{mk}}{\partial x_i \partial x_l} \right) \\ + \sum_{\varrho\sigma} \gamma_{\varrho\sigma} \left(\begin{bmatrix} im \\ \varrho \end{bmatrix} \begin{bmatrix} kl \\ \sigma \end{bmatrix} - \begin{bmatrix} il \\ \varrho \end{bmatrix} \begin{bmatrix} km \\ \sigma \end{bmatrix} \right). \end{aligned}$$

Durch kovariante algebraische und differentielle Operationen erhält man aus dem Riemannschen Differentialtensor und dem Diskriminantentensor (§ 3, Formel 38) das vollständige System der Differentialtensoren (also auch der Differentialinvarianten) der Mannigfaltigkeit.

---

1) Riemann, Ges. Werke, S. 270.
2) Christoffel, l. c., S. 54.

(ik, lm) heißen auch die **Christoffel**schen **Vier-Indizes-Symbole erster Art.** Von Bedeutung sind neben diesen die **Vier-Indizes-Symbole zweiter Art**

$$(44) \quad \{ik, lm\} = \frac{\partial \{^{il}_{\ k}\}}{\partial x_m} - \frac{\partial \{^{im}_{\ k}\}}{\partial x_l} + \sum_\varrho \left(\{^{\ il}_\varrho\}\{^{\varrho m}_{\ k}\} - \{^{\ im}_\varrho\}\{^{\varrho l}_{\ k}\} \right),$$

die mit jenen in der Beziehung stehen

$$(45) \quad \begin{cases} \{i\varrho, lm\} = \sum_k \gamma_{\varrho k}(ik, lm), \text{ oder aufgelöst} \\ (ik, lm) = \sum_\varrho g_{k\varrho}\{i\varrho, lm\}. \end{cases}$$

Den Vier-Indizes-Symbolen zweiter Art kommt in der allgemeinen Vektoranalysis die Bedeutung der Komponenten eines gemischten Tensors, kovariant vom dritten, kontravariant vom ersten Range zu.[1]) Die hervorragende Bedeutung dieser Begriffsbildungen für die Differentialgeometrie[2]) einer durch ihr Linienelement gegebenen Mannigfaltigkeit macht es a priori wahrscheinlich, daß diese allgemeinen Differentialtensoren auch für das Problem der Differentialgleichungen eines Gravitationsfeldes von Bedeutung sein dürften. Es gelingt in der Tat zunächst, einen kovarianten Differentialtensor zweiten Ranges und zweiter Ordnung G_{im} anzugeben, der in jene Gleichungen eintreten könnte, nämlich

$$(46) \quad G_{im} = \sum_{kl} \gamma_{kl}(ik, lm) = \sum_k \{ik, km\}.$$

Allein es zeigt sich, daß sich dieser Tensor im Spezialfall des unendlich schwachen statischen Schwerefeldes nicht auf den Ausdruck $\varDelta \varphi$ reduziert. Wir müssen daher die Frage offen lassen, inwiefern die allgemeine Theorie der mit einem Gravitationsfeld verknüpften Differentialtensoren mit dem Problem der Gravitationsgleichungen zusammenhängt. Ein solcher Zusammenhang müßte vorhanden sein, sofern die Gravitationsgleichungen beliebige Substitutionen zuzulassen hätten; allein in diesem Falle scheint es ausgeschlossen zu sein, Differentialgleichungen zweiter Ordnung aufzufinden. Würde dagegen feststehen, daß die Gravitationsgleichungen nur eine gewisse Gruppe von Transformationen gestatten, so wäre es verständlich, wenn man mit den von der allgemeinen Theorie gelieferten Differentialtensoren nicht auskommt. Wie im physikalischen Teile ausgeführt ist, sind wir nicht imstande, zu diesen Fragen Stellung zu nehmen. —

---

1) Es folgt dies aus der ersten der Gleichungen 45.

2) Das identische Verschwinden des Tensors R_{iklm}, stellt die notwendige und hinreichende Bedingung dafür dar, daß die Differentialform auf die Form $\sum_i dx_i'^2$ transformiert werden kann.

3. Zur Ableitung der Gravitationsgleichungen.

Die von Einstein beschriebene Herleitung der Gravitationsgleichungen (I. Teil, § 5), wird im Einzelnen folgendermaßen durchgeführt.

Wir gehen aus von dem in der Energiebilanz mit Gewißheit zu erwartenden Gliede

$$(47) \qquad U = \sum_{\alpha\beta\mu\nu} \frac{\partial g_{\mu\nu}}{\partial x_\sigma} \frac{\partial}{\partial x_\sigma} \left(\sqrt{g}\, \gamma_{\alpha\beta} \frac{\partial \gamma_{\mu\nu}}{\partial x_\beta} \right)$$

und formen durch partielle Integration um.[1] Es wird so

$$U = \sum_{\alpha\beta\mu\nu} \frac{\partial}{\partial x_\alpha} \left(\sqrt{g}\, \gamma_{\alpha\beta} \frac{\partial \gamma_{\mu\nu}}{\partial x_\beta} \frac{\partial g_{\mu\nu}}{\partial x_\sigma} \right) - \sum_{\alpha\beta\mu\nu} \sqrt{g}\, \gamma_{\alpha\beta} \frac{\partial \gamma_{\mu\nu}}{\partial x_\beta} \cdot \frac{\partial^2 g_{\mu\nu}}{\partial x_\sigma \partial x_\alpha}.$$

Die erste der auf der rechten Seite stehenden Summen hat die gewünschte Form einer Summe von Differentialquotienten und sei bezeichnet mit A, so daß

$$(48) \qquad A = \sum_{\alpha\beta\mu\nu} \frac{\partial}{\partial x_\alpha} \left(\sqrt{g}\, \gamma_{\alpha\beta} \frac{\partial \gamma_{\mu\nu}}{\partial x_\beta} \frac{\partial g_{\mu\nu}}{\partial x_\sigma} \right).$$

In der zweiten der rechtsstehenden Summen führen wir wieder partielle Integration aus. Dann lautet die Identität

$$U = A - \sum_{\alpha\beta\mu\nu} \frac{\partial}{\partial x_\sigma} \left(\sqrt{g} \cdot \gamma_{\alpha\beta} \frac{\partial \gamma_{\mu\nu}}{\partial x_\beta} \cdot \frac{\partial g_{\mu\nu}}{\partial x_\alpha} \right) + \sum_{\alpha\beta\mu\nu} \frac{\partial g_{\mu\nu}}{\partial x_\alpha} \cdot \frac{\partial}{\partial x_\sigma} \left(\sqrt{g} \cdot \gamma_{\alpha\beta} \frac{\partial \gamma_{\mu\nu}}{\partial x_\beta} \right).$$

Die erste der rechts entstandenen Summen kann als eine Summe von Differentialen geschrieben werden und möge mit

$$(49) \qquad B = \sum_{\alpha\beta\mu\nu} \frac{\partial}{\partial x_\sigma} \left(\sqrt{g}\, \gamma_{\alpha\beta} \frac{\partial \gamma_{\mu\nu}}{\partial x_\beta} \frac{\partial g_{\mu\nu}}{\partial x_\alpha} \right)$$

bezeichnet sein. In der zweiten Summe differentiieren wir aus. Dann wird

$$U = A - B + \sum_{\alpha\beta\mu\nu} \frac{\partial g_{\mu\nu}}{\partial x_\alpha} \left(\gamma_{\alpha\beta} \frac{\partial \gamma_{\mu\nu}}{\partial x_\beta} \frac{\partial \sqrt{g}}{\partial x_\sigma} + \sqrt{g} \cdot \frac{\partial \gamma_{\mu\nu}}{\partial x_\beta} \cdot \frac{\partial \gamma_{\alpha\beta}}{\partial x_\sigma} + \sqrt{g} \cdot \gamma_{\alpha\beta} \frac{\partial^2 \gamma_{\mu\nu}}{\partial x_\beta \partial x_\sigma} \right),$$

oder wenn man im zweiten Summanden die Formel (29) des § 2 anwendet und im dritten Summanden partiell integriert

$$U = A - B + \sum_{\alpha\beta\mu\nu ik} \gamma_{\alpha\beta} \frac{\partial g_{\mu\nu}}{\partial x_\alpha} \frac{\partial \gamma_{\mu\nu}}{\partial x_\beta} \cdot \frac{\sqrt{g}}{2} \gamma_{ik} \frac{\partial g_{ik}}{\partial x_\sigma} - \sum_{\alpha\beta\mu\nu ik} \sqrt{g} \cdot \frac{\partial g_{\mu\nu}}{\partial x_\alpha} \cdot \frac{\partial \gamma_{\mu\nu}}{\partial x_\beta} \cdot \gamma_{\alpha i} \gamma_{\beta k} \frac{\partial g_{ik}}{\partial x_\sigma}$$
$$+ \sum_{\alpha\beta\mu\nu} \frac{\partial}{\partial x_\beta} \left(\sqrt{g}\, \gamma_{\alpha\beta} \frac{\partial g_{\mu\nu}}{\partial x_\alpha} \cdot \frac{\partial \gamma_{\mu\nu}}{\partial x_\sigma} \right) - \sum_{\alpha\beta\mu\nu} \frac{\partial \gamma_{\mu\nu}}{\partial x_\sigma} \frac{\partial}{\partial x_\beta} \left(\sqrt{g}\, \gamma_{\alpha\beta} \frac{\partial g_{\mu\nu}}{\partial x_\alpha} \right).$$

---

[1] Die Herleitung der gesuchten Identität vereinfacht sich, wenn wir den Faktor $\sqrt{g}$ unter das Differentiationszeichen setzen, ohne daß das Resultat hiervon abhängig wäre.

Die beiden ersten Summen haben die Form von Gliedern, wie wir sie auf die linke Seite unserer Identität setzen. Wir bezeichnen sie mit

$$(50) \qquad V = \tfrac{1}{2} \sum_{\alpha\beta\mu\nu ik} \frac{\partial g_{ik}}{\partial x_\alpha} \cdot \sqrt{g} \cdot \gamma_{\alpha\beta} \gamma_{ik} \frac{\partial g_{\mu\nu}}{\partial x_\alpha} \frac{\partial \gamma_{\mu\nu}}{\partial x_\beta}$$

$$(51) \qquad W = \sum_{\alpha\beta\mu\nu ik} \frac{\partial g_{ik}}{\partial x_\alpha} \cdot \sqrt{g} \cdot \gamma_{\alpha i} \gamma_{\beta k} \frac{\partial g_{\mu\nu}}{\partial x_\alpha} \cdot \frac{\partial \gamma_{\mu\nu}}{\partial x_\beta}.$$

Die dritte der rechts stehenden Summen hat die Form einer Summe von Differentialquotienten; eliminiert man in ihr $\dfrac{\partial \gamma_{\mu\nu}}{\partial x_\alpha}$ vermöge jener Formel (29), so erweist sie sich als die schon eingeführte Größe A. In der letzten Summe endlich ersetzen wir nach der gleichen Formel $\dfrac{\partial \gamma_{\mu\nu}}{\partial x_\alpha}$. Wir finden so

$$U - V + W = 2A - B + \sum_{\alpha\beta\mu\nu ik} \gamma_{\mu i} \gamma_{\nu k} \frac{\partial g_{ik}}{\partial x_\alpha} \frac{\partial}{\partial x_\beta} \left(\sqrt{g} \, \gamma_{\alpha\beta} \frac{\partial g_{\mu\nu}}{\partial x_\alpha} \right),$$

oder

$$U - V + W = 2A - B + \sum_{\alpha\beta\mu\nu ik} \frac{\partial g_{ik}}{\partial x_\alpha} \cdot \frac{\partial}{\partial x_\beta} \left(\sqrt{g} \cdot \gamma_{\alpha\beta} \gamma_{\mu i} \gamma_{\nu k} \frac{\partial g_{\mu\nu}}{\partial x_\alpha} \right)$$

$$- \sum_{\alpha\beta\mu\nu ik} \frac{\partial g_{ik}}{\partial x_\alpha} \frac{\partial g_{\mu\nu}}{\partial x_\alpha} \sqrt{g} \cdot \gamma_{\alpha\beta} \frac{\partial}{\partial x_\beta} (\gamma_{\mu i} \gamma_{\nu k}).$$

Die erste dieser Summen wird wegen (29), d. h. wegen

$$\sum_{\mu\nu} \gamma_{i\mu} \gamma_{\nu k} \frac{\partial g_{\mu\nu}}{\partial x_\alpha} = - \frac{\partial \gamma_{ik}}{\partial x_\alpha}$$

zu

$$- \sum_{\alpha\beta ik} \frac{\partial g_{ik}}{\partial x_\alpha} \frac{\partial}{\partial x_\beta} \left(\sqrt{g} \gamma_{\alpha\beta} \frac{\partial \gamma_{ik}}{\partial x_\alpha} \right) = - U.$$

Die zweite können wir, wegen der Vertauschbarkeit von i und k, μ und ν, schreiben als

$$2X = 2 \cdot \sum_{\alpha\beta\mu\nu ik} \frac{\partial g_{ik}}{\partial x_\alpha} \cdot \sqrt{g} \cdot \gamma_{\alpha\beta} \gamma_{\mu i} \frac{\partial g_{\mu\nu}}{\partial x_\alpha} \cdot \frac{\partial \gamma_{\nu k}}{\partial x_\beta}$$

$$= - 2 \cdot \sum_{\alpha\beta\mu\nu ik} \frac{\partial g_{ik}}{\partial x_\alpha} \cdot \sqrt{g} \, \gamma_{\alpha\beta} g_{\mu\nu} \frac{\partial \gamma_{i\mu}}{\partial x_\alpha} \frac{\partial \gamma_{k\nu}}{\partial x_\beta}.$$

Die gesuchte Identität lautet also

$$2U - V + W + 2X = 2A - B,$$

ist also identisch der im I. Teil, § 5 gegebenen.

Joint Papers by A. Einstein and M. Grossmann

Einstein, Albert, und Grossmann, Marcel (1913): *"Entwurf einer verallgemeinerten Relativitätstheorie und einer Theorie der Gravitation"* [Outline of a Generalised Theory of Relativity and of a Theory of Gravitation], *Zeitschrift für Mathematik und Physik* **62**, No. 3, pp. 225–261; CPAE, Vol. 4, Doc. 13.
Einstein, Albert and Grossmann, Marcel (1914): *"Kovarianzeigenschaften der Feldgleichungen der auf die verallgemeinerte Relativitatstheorie gegründeten Gravitationstheorie"* ["Covariance Properties of the Field Equations of a Theory of Gravitation based upon the Generalised Theory of Relativity"], *Zeitschrift für Mathematik und Physik*, May 29th, 1914; CPAE, Vol. 6, Doc. 2.

© Springer International Publishing AG, part of Springer Nature 2018
C. Graf-Grossmann, *Marcel Grossmann*, Springer Biographies,
https://doi.org/10.1007/978-3-319-90077-3

Literature

Publications by Marcel Grossmann (cf. also Backnotes)

Über die metrischen Eigenschaften kollinearer Gebilde, 1902, Huber & Co.,
   Frauenfeld.
*Die Konstruktion des geradlinigen Dreiecks der nichteuklidischen Geometrie aus
   den drei Winkeln*, 1904, *Mathematische Annalen*, Vol. 58, pp. 578–582.
Die fundamentalen Konstruktionen der nichteuklidischen Geometrie, 1904, in:
   *Beilage zum Programm der Thurgauischen Kantonsschule für das
Schuljahr* 1903/04 [Supplement to the Programme of the Canton School for
   the 1903/04 School Year], Huber & Co., Frauenfeld.
Metrische Eigenschaften reziproker Bündel, 1905, *Archiv für Mathematik und
   Physik*, Vol. 9, pp. 143–150.
Darstellende Geometrie. Leitfaden für den Unterricht an höheren Lehranstalten,
   1906, Helbing & Lichtenhahn, Basel.
Analytische Geometrie. Leitfaden für den Unterricht an höheren Lehranstalten,
   1906, Helbing & Lichtenhahn, Basel.
Einleitendes Referat, 1909, in: *Jahrbuch des Vereins Schweizerischer Gymna-
   siallehrer*, Vol. 39, pp. 20–24.
Projektive Konstruktionen in der hyperbolischen Geometrie, 1909, *Mathema-
   tische Annalen*, Vol. 68, pp. 141–144.
Über den Aufbau der Geometrie, 1909, in: *Schweizerische Pädagogische
   Zeitschrift*, Vol. 19 (V), pp. 282–297.
Lösung eines geometrischen Problems der Photogrammetrie, 1910, in: *Schweizerische
   Naturforschende Gesellschaft, Verhandlungen*, Vol. 93, pp. 338–339.

© Springer International Publishing AG, part of Springer Nature 2018   **249**
C. Graf-Grossmann, *Marcel Grossmann*, Springer Biographies,
https://doi.org/10.1007/978-3-319-90077-3

Der mathematische Unterricht an der Eidgenössischen Technischen Hochschule, 1911, Georg & Co., Basel and Geneva.

Projektiver Beweis der absoluten Parallelenkonstruktion von Lobatschefskij, 1912, in: *Schweizerische Naturforschende Gesellschaft, Verhandlungen,* Vol. 95, pp. 130–133.

Einführung in die darstellende Geometrie. Leitfaden für den Unterricht an höheren Lehranstalten, 1912, Helbing & Lichtenhahn, Basel.

Über nichteuklidische Geometrie, 1912, in: *Naturforschende Gesellschaft Zürich, Vierteljahrsschrift,* Vol. 57, pp. XVI–XXIV.

Entwurf einer verallgemeinerten Relativitätstheorie und einer Theorie der Gravitation. (Physical Part by A. Einstein, Mathematics Part by von M. Grossmann), 1913, *Zeitschrift für Mathematik und Physik,* Vol. 62 (3), pp. 225–261. Published as a separate reprint by Teubner, Leipzig and Berlin.

Die Vorbildung der Kandidaten für die Technische Hochschule, 1913, *Jahrbuch des Vereins Schweizerischer Gymnasiallehrer,* Vol. 43, pp. 129–134.

Prof. Dr. Otto Wilhelm Fiedler (1832–1912), 1913, *Schweizerische Naturforschende Gesellschaft, Verhandlungen,* Vol. 96, pp. 20–27.

Die Zentralprojektion in der absoluten Geometrie, 1913, in: *Proceedings of the Fifth International Congress of Mathematicians* (Cambridge, 22–28 August 1912), Cambridge University Press, Cambridge, pp. 66–69.

Mathematische Begriffsbildungen, Methoden und Probleme zur Gravitationstheorie, 1913, *Schweizerische Naturforschende Gesellschaft, Verhandlungen,* Vol. 96, pp. 138–140.

Mathematische Begriffsbildungen zur Gravitationstheorie, 1913, *Naturforschende Gesellschaft Zürich, Vierteljahrsschrift,* Vol. 58, pp. 291–297.

(with A. Einstein) *Kovarianzeigenschaften der Feldgleichungen der auf die verallgemeinerte Relativitätstheorie gegründeten Gravitationstheorie,* 1914, *Zeitschrift für Mathematik und Physik,* Vol. 63, pp. 215–225.

Définitions, Méthodes et Problèmes mathématiques relatifs á la Théorie de la Gravitation, 1914, *Archives des sciences physiques et naturelles,* Vol. 37, pp. 13–19.

Darstellende Geometrie, 1915, Teubner, Leipzig und Berlin.

Anregungen zum Problem der nationalen Erziehung, 1915, in: *Jahrbuch des Vereins Schweizerischer Gymnasiallehrer,* Vol. 44, pp. 1–12.

Nationale Forderungen an die schweizerische Mittelschule, 1915, Rascher & Cie., Zurich.

Nationale Erneuerung und nationale Erziehung I, 1915, in: *Neue Zürcher Zeitung,* Zurich.

Nationale Erneuerung und nationale Erziehung II, 1915, in: *Neue Zürcher Zeitung,* Zurich.

Elemente der darstellenden Geometrie, 1917, Teubner, Berlin.

Einführung in die Darstellende Geometrie. Leitfaden für den Unterricht an höheren Lehranstalten, 1917, Helbing & Lichtenhahn, Basel.

Über die Rolle der Frau in der nationalen Erziehung unserer Jugend, 1917, Zürcher & Furrer, Zurich.

Unsere Presse, 1918, in: *Neue Schweizer Zeitung*, No. 1 (2), Zurich.

Zur Mittelschulreform, 1919, in: *Schweizerische Bauzeitung*, No. 74 (22), S. 268–270.

Einstein-Hetze in Berlin, 1920, in: *Neue Schweizer Zeitung*, No. 2 (70), Zurich.

Aus unserer Presse. Physikprofessoren, 1920, in: *Neue Schweizer Zeitung*, No. 2 (47), Zurich.

Ein neues Weltbild, 1920, in: *Neue Schweizer Zeitung*, No. 2 (1/2), Zurich.

Mise au point mathématique, 1920, Archives des sciences physiques et naturelles, Vol. 2, pp. 497–499.

Darstellende Geometrie II. Teil, 1921, Teubner, Leipzig und Berlin.

Eidgenössische Maturitätsreform, 1921, in: *Schweizerische Bauzeitung*, No. 78 (14), pp. 167–169.

Darstellende Geometrie I. Teil, 1922, Teubner, Berlin.

Prof. Rudolf Escher (1848–1921), 1922, in: *Schweizerische Naturforschende Gesellschaft, Verhandlungen*, Vol. 103, pp. 13–14.

Elliptische Geometrie im Antipolarsystem, 1922, in: *Schweizerische Naturforschende Gesellschaft, Verhandlungen*, Vol. 103, p. 171.

Eidgenössische Maturitätsreform, 1st Part, 1922, *Schweizerische Pädagogische Zeitschrift*, Vol. 32 (I), pp. 33–39.

Eidgenössische Maturitätsreform, 2nd Part, 1922, *Schweizerische Pädagogische Zeitschrift*, Vol. 32 (II), p. 1–8.

Reifeerklärung, 1923, Fretz, Zurich.

Uneigentliche Raumelemente, 1923, Schweizerische Pädagogische Zeitschrift, Vol. 33 (II), pp. 60–62.

Die Lösung des Maturitätskonflikts, 1923, in: *Neue Zürcher Zeitung*, Zurich.

Darstellung des Horopters, 1925, Naturforschende Gesellschaft Zürich, Vierteljahrsschrift, Vol. 70, pp. 66–76.

Das vollständige Fokalsystem einer ebenen algebraischen Kurve, 1925, Acta Litterarum ac Scientiarum (Szeged), Sectio Scientarum Mathematicarum, Vol. 2(3), pp. 178–181.

Diagnosen, 1926, Schweizerische Pädagogische Zeitschrift, Vol. 36 (III), pp. 65–72.

Darstellende Geometrie für Maschineningenieure, 1927, Springer, Berlin.

Präzisions-Schlagexcenter für mechanische Webstühle, geometrische Formgebung und zwangsläufige Herstellung, 1927, Schweizerische Bauzeitung, Vol. 90 (22), pp. 279–282.

Fachbildung, Geisteskultur und Phantasie, in: *Festschrift* Prof. Dr. A. Stodola *zum 70. Geburtstag*, 1929, Orell Füssli, Zurich und Leipzig, pp. 187–190.

Darstellung des Kreises und der Kegelschnitte, 1930, in *Commentarii mathematici helvetici*, Vol. 2, pp. 174–177.

Fernparallelismus? Richtigstellung der gewählten Grundlagen für eine einheitliche Feldtheorie, 1931, *Naturforschende Gesellschaft Zürich, Vierteljahrsschrift*, Vol. 76, pp. 42–60.

Darstellende Geometrie. Abbildungsverfahren, Kurven und Flächen, 1932, Teubner, Leipzig und Berlin.

Publications about Marcel Grossmann and his Circle

Baumann, Walter: Zürich, *La Belle Epoque*, 1973, Orell Füssli Verlag, Zürich.

Baumann, Walter: Cattani, Alfred, Loetscher, Hugo und Scheidegger, Ernst, *Zürich zurückgeblättert 1870–1914, Werden und Wandel einer Stadt*, 1979, Verlag Neue Zürcher Zeitung, Zürich.

Bianchi, Luigi: *Vorlesungen über Differentialgeometrie (autorisierte deutsche Übersetzung von Max Lukat)*, 1899, Teubner, Leipzig.

Bonola, Roberto: *Die Nichteuklidische Geometrie. Historisch-kritische Darstellung ihrer Entwicklung*, 1908, Teubner, Leipzig.

Cabanes, Bruno und Dumeuil, Anne: *Der Erste Weltkrieg, eine europäische Katastrophe*, 2013, Konrad Theiss Verlag, Berlin.

Christoffel, E.B.: *Über die Transformation der homogenen Differentialausdrücke zweiten Grades*, 1869, *Journal für die reine und angewandte Mathematik*, **70**, No. 1, pp. 46–70.

Clark, Christopher: *Die Schlafwandler. Wie Europa in den Ersten Weltkrieg zog*, 2013, Deutsche Verlags-Anstalt, München.

Einstein, Albert: *Physikalische Grundlagen einer Gravitationstheorie*, 1913, *Naturforschende Gesellschaft Zürich, Vierteljahrsschrift*, **58**, pp. 284–290.

Einstein, Albert: *Zur allgemeinen Relativitätstheorie*, 1915, *Königlich Preußische Akademie der Wissenschaften, Sitzungsberichte* 1915, pp. 778–786.

Einstein, Albert: *Riemann-Geometrie mit Aufrechterhaltung des Begriffes des Fernparallelismus*, 1928, *Preußische Akademie der Wissenschaften, Berlin, physikalisch-mathematische Klasse, Sitzungsberichte* 1928, pp. 217–221.

Einstein, Albert: *Relativity. The Special and the General Theory*, 100th Anniversary Edition, With Commentary and Background Material by Hanoch Gutfreund and Jürgen Renn, 2015, Princeton University Press, Princeton.

Falconnier, René: *Die Chronik derer zu Lichtenhayn*, 1979, G. Krebs AG, Basel.

Fiedler, Otto Wilhelm: *Die darstellende Geometrie in organischer Verbindung mit der Geometrie der Lage*, I. Theil, Die Methoden der darstellenden und die Elemente der projectivischen Geometrie, 1883, B.G. Teubner, Leipzig.

Fiedler, Otto Wilhelm: *Die darstellende Geometrie in organischer Verbindung mit der Geometrie der Lage*, II. Theil, Die darstellende Geometrie der krummen Linien und Flächen, 1885, B.G. Teubner, Leipzig.

Fiedler, Otto Wilhelm: *Die darstellende Geometrie in organischer Verbindung mit der Geometrie der Lage*, III. Theil, Die constituirende und analytische Geometrie der Lage, 1888, B.G. Teubner, Leipzig.

Frei, Günther und Stammbach, Urs: *Hermann Weyl und die Mathematik an der ETH Zürich 1913–1930*, 1992, Birkhäuser, Basel, Boston, Berlin.

Frei, Günther und Stammbach, Urs: *Die Mathematiker an den Zürcher Hochschulen, 1994*, Birkhäuser, Basel, Boston, Berlin.

Friedrich, Jörg: *14/18, Der Weg nach Versailles*, 2014, Propyläen Verlag, Berlin.

Geiser, Carl Friedrich: *Zur Erinnerung an Theodor Reye*, 1921, *Naturforschende Gesellschaft Zürich, Vierteljahrsschrift*, **66**, pp. 158–180.

Giulini, Domenico: *Spezielle Relativitätstheorie*, 2004, Fischer Taschenbuch Verlag, Frankfurt/Main.

Giulini, Domenico und Straumann, Norbert: *Einstein's Impact on the Physics of the Twentieth Century*, 2006, in: Studies in the History and Philosophy of Modern Physics, **37**, pp. 115–173.

Guanzini, Catherine und Wegelin, Peter: *Kritischer Patriotismus*, 1989, Paul Haupt, Bern.

Gutfreund, Hanoch und Renn, Jürgen: *The Road to Relativity*, 2015, Princeton University Press, Princeton.

Helmholtz, Hermann von: *Wissenschaftliche Abhandlungen*, 1888, Vol. 2, Barth, Leipzig.

Hermann, Robert: *Ricci and Levi-Civita's Tensor Analysis Paper*, Translation, Comments, and Additional Material, 1975, Math Sci Press, Brookline, Massachusetts.

Isaacson, Walter: *Einstein, His Life and Universe*, 2007, Simon and Schuster, New York.

Janssen, Michel und Lehner, Christoph (Eds.): *The Cambridge Companion to Einstein*, 2014, Cambridge University Press, Cambridge.

Karl Dändliker: *Darstellende hyperbolische Geometrie*, Dissertation, 1919, ETH Zürich.

Katzarova, Panka: *Analoga der Steinerschen Konstruktion in einer absoluten Geometrie*, Ph.D. Dissertation, 1977, Ruhr-Universität Bochum.

Katzarova, *Panka: Analoga der Steinerschen Konstruktion in einer absoluten Geometrie*, 1981, *Rendiconti di Circolo Matematico di Palermo*, **30**, pp. 287–299.

Kollros, Louis: *Géométrie descriptive*, 1918, Orell Füssli, Zürich.

Kollros, Louis: *Prof. Dr. Marcel Grossmann 1878–1936*, 1937, *Schweizerische Naturforschende Gesellschaft, Verhandlungen*, **118**, pp. 325–329.

Kreis, Georg: *Insel der unsicheren Geborgenheit, Die Schweiz in den Kriegsjahren 1914–1918*, 2014, Verlag Neue Zürcher Zeitung, Zürich.

Leutenegger, Emil: *Über Kegelschnitte in der hyperbolischen Geometrie*, Dissertation, 1923, ETH Zürich.

Liebmann, Heinrich: *Nichteuklidische Geometrie*, 1905, Göschen, Leipzig.

Liebmann, Heinrich: *Die elementaren Konstruktionen der nichteuklidischen Geometrie*, 1911, *Jahresbericht der Deutschen Mathematiker-Vereinigung*, **20**, pp. 56–69.

Melinz, Gerhard und Zimmermann, Susan: *Wien, Prag, Budapest, Blütezeit der Habsburgermetropolen*, 1996, Promedia, Wien.

Mettler, Ernst: *Anwendung der stereographischen Projektion auf Konstruktionen im nicht-euklidischen Raum*, Dissertation, 1916, ETH Zürich.

Neuenschwander, Erwin: *100 Jahre Schweizerische Mathematische Gesellschaft 1910–2010*, 2010, pp. 23–106.

N. N.: *Budapest Then & Now (Budapest Anno Es Most)*, 2012, Kiscelli Muzeum, Budapest.

Reich, Karin: *Die Entwicklung des Tensorkalküls. Vom absoluten Differentialkalkül zur Relativitätstheorie*, 1994, Birkhäuser, Basel.

Renn, Jürgen (Ed.): *The Genesis of General Relativity* (4 vols.), 2007, Renn, Jürgen, Springer, Dordrecht.

Renn, Jürgen und Sauer, Tilman: *Einstein's Züricher Notizbuch*, 1996, *Physikalische Blätter*, **52**, No. 9, pp. 865–872.

Renn, Jürgen und Sauer, Tilman: *Heuristics and Mathematical Representation in Einstein's Search for a Gravitational Field Equation*, 1999, in: The Expanding Worlds of General Relativity, pp. 87–125.

Ricci, Gregorio und Levi-Civita, Tullio: *Méthodes de calcul différentiel absolu et leurs applications*, 1901, *Mathematische Annalen*, **54**, pp. 125–201.

Riemann, Bernhard: *Commentatio mathematica, qua respondere tentatur quaestioni ab Illma Academia Parisieni propositae*, 1876, in: *Gesammelte Mathematische Werke und wissenschaftlicher Nachlass*, pp. 370–383.

Rogger, Franziska: *Einsteins Schwester, Maja Einstein – ihr Leben und ihr Bruder Albert*, 2005, Verlag Neue Zürcher Zeitung, Zürich.

Rosenfeld, B.A.: *A History of Non-Euclidean Geometry, Evolution of the Concept of a Geometric Space*, 1988, Springer, New York.

Rosenkranz, Ze'ev: *Albert Einstein, Privat und ganz persönlich*, 2004, Verlag Neue Zürcher Zeitung, Zürich.

Salmon, George: *Analytische Geometrie des Raumes (deutsch bearbeitet von W. Fiedler)*, I. Theil, Die Elemente und die Theorie der Flächen zweiten Grades, 1898, Teubner, Leipzig.

Salmon, George: *Analytische Geometrie des Raumes (deutsch bearbeitet von W. Fiedler)*, II. Theil, Analytische Geometrie der Curven im Raume und der algebraischen Flächen, 1898, Teubner, Leipzig.

Sauer, Tilman: *Einstein's Review Paper on General Relativity*, 2005, in: Landmark Writings of Western Mathematics, 1640–1940, pp. 802–822.

Sauer, Tilman: *Field equations in teleparallel space-time, Einstein's Fernparallelismus approach toward unified field theory*, 2006, in: *Historia Mathematica*, **33**, pp. 399–439.

Sauer, Tilman: *Einstein's Unified Field Theory Program*, 2007, in: The Cambridge Companion to Einstein, pp. 281–305.

Sauer, Tilman: *Marcel Grossmann and his contribution to the general theory of relativity*, 2015, in: Proceedings of the 13th Marcel Grossmann Meeting on Recent Developments in Theoretical and Experimental General Relativity, Gravitation, and Relativistic Field Theory, pp. 456–503.

Saxer, Walter: *Marcel Grossmann (1878–1936; Mitglied der Gesellschaft seit 1908)*, 1936, in: *Naturforschende Gesellschaft Zürich, Vierteljahrsschrift*, **81**, pp. 322–326.

Schilling, Martin: *Catalog mathematischer Modelle für den höheren mathematischen Unterricht*, 1911, Verlag von Martin Schilling, Leipzig.

Schuh, Fred: *Die Horopterkurve*, 1902, *Zeitschrift für Mathematik und Physik*, **47**, pp. 375–399.

Schulmann, Robert (Ed.): *Seelenverwandte. Der Briefwechsel zwischen Albert Einstein und Heinrich Zangger*, 2012, Schulmann, Robert, Zurich, Verlag Neue Zürcher Zeitung, Zürich.

Schwarzenbach, Alexis: *Das verschmähte Genie. Albert Einstein und die Schweiz*, 2005, Deutsche Verlags-Anstalt, München.

Seelig, Carl: *Albert Einstein und die Schweiz*, 1952, Europa-Verlag, Zürich, Stuttgart, Wien.

Stammbach, Urs: *Sternstunden der Mathematikgeschichte*, 2013, Selbstverlag, Zürich.

Straumann, Norbert: *Einstein's Zurich Notebook and his journey to general relativity*, 2011, *Annalen der Physik*, **523**, No. 6, pp. 488–500.

Straumann, Norbert: *On the Einstein-Grossmann Collaboration 100 Years ago*, 2013, in: Swiss Physical Society, *Mitteilungen*, **40**, pp. 48–51.

Trbuhovic-Gjuric, Desanka: *Im Schatten Albert Einsteins. Das tragische Leben der Mileva Einstein-Marić*, 1983, Paul Haupt, Bern/Stuttgart.

Ulshöfer, Klaus und Tilp, Dietrich: *Darstellende Geometrie in systematischen Beispielen*, 2011, C.C. Buchners Verlag, Bamberg.

Vaterlaus, Ernst: *Konstruktionen in der Bildebene der hyperbolischen Zentralprojektion*, Dissertation, 1916, ETH Zürich.

Vonlanthen, Adolf; Lattmann, Urs Peter und Egger, Eugen: *Maturität und Gymnasium*, 1978, Paul Haupt, Bern.

Zacharias, Max: *Elementargeometrie und elementare nicht-euklidische Geometrie in synthetischer Behandlung*, 1913, in: *Encyclopädie der mathematischen Wissenschaften mit Einschluss ihrer Anwendungen*, Vol. 3, Geometrie (1914–1931), Erster Theil, zweite Hälfte, No. 9, pp. 859–1172.

A recommendation for interested readers:

The Collected Papers of Albert Einstein (CPAE), which are now available online (www.einstein.caltech.edu, http://einsteinpapers.press.princeton.edu), as well as the digitised documents of Marcel Grossmann (www.e-manuscripta.ch/zuz).

Picture Credits

All the photos are from the private possession of the Grossmann family with the exceptions of:

Cover jacket, front, pp. 1, 112: Portrait of Marcel Grossmann: 1906, ETH Library, Image archives, Port_01239.

Inside front cover: Graphics, Saskia Noll.

Inside back cover: ETH-Bibliothek Zurich, Hs 421:1.

p. 17.	Advertisement for *Japy Frères*: www.clockguy.com.
p. 48.	Contract, Grossmann und Rauschenbach Co., Municipal Archives Schaffhausen, D IV, 01.06.07/II.
p. 55.	Balance sheet of Grossmann und Rauschenbach, Municipal Archives Schaffhausen, D IV, 01.06.07/II.
p. 92.	Postcard from Marcel Grossmann to Pauline Einstein, ETH Library, University Archives, Hs 421a:19.
p. 10.3	Doctoral certificate of Marcel Grossmann, ETH Library, University Archives, Hs 421a:18.
p. 176.	*Bernisches Historisches Museum*, Inventory No. 62783.
p. 179ff.	Einstein Archive, No. II-464.
p. 231,	above. Image archive, Tilman Sauer.

© Springer International Publishing AG, part of Springer Nature 2018
C. Graf-Grossmann, *Marcel Grossmann*, Springer Biographies,
https://doi.org/10.1007/978-3-319-90077-3

p. 231,   below. Illustration from Grossmann, Marcel: *Otto Wilhelm Fielder* (1832–1912), in: Verhandlungen der Schweizerischen Naturforschenden Gesellschaft **96**, 1913, p. 20.

p. 235.   ETH Library, University Archives, Hs 421:11, Sketch p. 16.

p. 239.   from Marcel Grossmann, "Die Konstruktion des gradlinigen Dreiecks der nichteuklidischen Geometrie aus den drei Winkeln", Mathematische Annalen 58, 1904, p. 579.

p. 241.   from Marcel Grossmann, "Einführung in die Darstellende Geometrie. Leitfaden für den Unterricht an höheren Lehranstalten". 3rd edition. Basel: Helbig & Lichtenhahn, 1917. Title picture.

p. 247,   above. Albert Einstein Archive, Sign. 3-006, p. 14L, © Hebrew University of Jerusalem.

p. 247,   below. Albert Einstein Archive, Sign. 3-006, S. 22R, © Hebrew University of Jerusalem.

p. 253.   © Tilman Sauer.

p. 256.   left. © Tilman Sauer.

p. 256,   right. from Marcel Grossmann, "*Darstellende Geometrie für Maschineningenieure*", Berlin 1927, p. 229.

All quotations and reproductions of works by Albert Einstein:
With the kind permission of the Einstein Archive, Hebrew University

Author Index

---

* The asterisks denote members of the Grossmann family

© Springer International Publishing AG, part of Springer Nature 2018
C. Graf-Grossmann, *Marcel Grossmann*, Springer Biographies,
https://doi.org/10.1007/978-3-319-90077-3

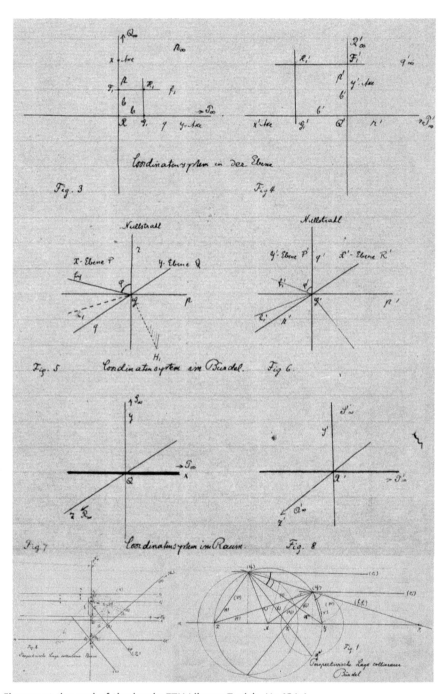

Figures at the end of the book: ETH Library Zurich, Hs 421:1

CPSIA information can be obtained
at www.ICGtesting.com
Printed in the USA
LVHW04*1914130618
580611LV00009B/268/P